What to Expect

Have you always wanted to learn how to solve logic grid puzzles, but they seemed too difficult? Or have you tried some, enjoyed them, but then got stuck when you got to the more complex clues? In that case, this book's for you!

We'll start with a sample 2x2 grid that shows what a logic grid looks like, in case you've never solved one before. This sample comes with a step-by-step solution, demonstrating the deduction process in detail. Following that, you'll find common examples of clues used in logic grid puzzles and the methods to handle each type of clue. These tips are provided alongside five 3x3 puzzles of increasing difficulty, solved step-by-step, with full explanations of the reasoning that leads to the solution. The puzzles are aligned with descriptions of the types of clues you'll see when solving logic grid puzzles. We encourage you to try solving each puzzle on your own. Only refer to the walkthrough if you face issues, and to check your work at the end.

Some of the explanations and examples are fairly basic, to accommodate readers with any level of logic grid experience. If you don't need every section of this guide, that's okay! Jump ahead to the section that's right for you.

This book also includes 25 puzzles to solve on your own, and solutions for each at the end of the book. After you've finished learning everything you need to know to solve these puzzles, test your skills and see if you can complete them all!

We hope you enjoy this comprehensive guide to logic grid puzzles. Happy solving!

Basic
Example

Conditions

Five children are celebrating birthdays in consecutive months.
1. Francine's birthday is in August.
2. Jason's birthday is in October, and he's a year older than Celeste.
3. The child turning 14 wasn't born in June.
4. The children celebrating birthdays in June and August were born in consecutive years, in no specific order.
5. Jason is one year younger than the person born in July.
6. Peter is one year older than Emily.

		Age					Month				
The Birthdays		15	14	13	12	11	October	September	August	July	June
Name	Emily										
	Peter										
	Jason										
	Francine										
	Celeste										
Month	June										
	July										
	August										
	September										
	October										

Initial Elimination

We can begin by marking the grid with any information given to us in the clues.

In this case, Clues 1 and 2 give us the birthdays of Jason and Francine. Since each child has only one possible birth month, we can eliminate all other months for Jason and Francine. Moreover, because each child's birthday is a different month, we can also eliminate all other children's names for August and October.

Clue 3 allows us to eliminate another possibility: We know the child turning 14 was not born in June, so we should mark that off as well.

	The Birthdays	Age					Month				
		15	14	13	12	11	October	September	August	July	June
Name	Emily						✗		✗		
	Peter						✗		✗		
	Jason						✓	✗	✗	✗	✗
	Francine						✗	✗	✓	✗	✗
	Celeste						✗		✗		
Month	June		✗								
	July										
	August										
	September										
	October										

Deduction

We know from Clue 6 that Peter is one year older than Emily, so Peter cannot be the youngest child, and Emily cannot be the oldest. Hence, we eliminate the youngest age for Peter, and the oldest age for Emily. Similarly (from Clue 2), Jason cannot be the youngest and Celeste cannot be the oldest. Clue 5 also uses a similar line of logic, allowing us to deduce that Jason cannot be the oldest and the person with the July birthday cannot be the youngest. Also, we know that Jason is older than Celeste but younger than the person born in July, which means Celeste cannot have been born in July.

Clue 4 tells us the children born in June and August were born in consecutive years. We know Francine was born in August, from Clue 1. If she was 15 years old, it would imply the person born in June is 14. However, Clue 3 clearly states that the child turning 14 wasn't born in June, which means it's impossible for Francine to be 15. Therefore, we can deduce that the only option for age 15 is Peter.

		Age					Month				
		15	14	13	12	11	October	September	August	July	June
Name	Emily	X					X		X		
	Peter	✓			X	X		X			
	Jason	X				X		X	✓	X	X
	Francine	X						X	X	X	X
	Celeste	X					X		X	X	
Month	June		X								
	July					X					
	August										
	September										
	October										

7

With each age we confirm, we eliminate the other possibilities within the corresponding column and row, allowing us to eliminate the other ages for Peter.

Since Emily is one year younger than Peter, we now know Emily's age as well: age 14. Consequently, Clue 3 tells us Emily couldn't have been born in June.

Additionally, if Peter were born in June, Emily and Francine would be the same age, because Clue 4 tells us Francine and Peter were born in consecutive years, and we've already determined Peter is the oldest. It's impossible for two children to share the same age; thus, Peter couldn't have been born in June. That leaves us with only one possibility for the child born in June: Celeste.

We know that Jason is one year younger than the person born in July from Clue 5. This eliminates July as a possibility for Peter because, if Peter were born in July, both Jason and Emily would be one year younger than Peter—the same age. As a result, the only possibility for July is Emily, which leaves September as the only remaining option for Peter.

And the puzzle is solved!

		Age					Month				
The Birthdays		15	14	13	12	11	October	September	August	July	June
Name	Emily	X	✓	X	X	X	X	X	X	✓	X
	Peter	✓	X	X	X	X	X	✓	X	X	X
	Jason	X	X	✓	X	X	✓	X	X	X	X
	Francine	X	X	X	✓	X	X	X	✓	X	X
	Celeste	X	X	X	X	✓	X	X	X	X	✓
Month	June	X									
	July				X						
	August										
	September										
	October										

Solving with Explanations

Basic Cross-Referencing

Solving a logic grid puzzle requires matching items from different categories, meaning it's always worthwhile to check for connections between the puzzle's categories.

For example, let's assume we have four toys made of wood, plastic, paper, and rubber. We'll also assume they're colored red, yellow, green, and blue, and are shaped like a cube, a cone, a ball, and a star.

Let's look at the following conditions:

- **The cone is blue.**
- **The cone is made of plastic.**

This clearly means the blue toy is made of plastic, since the attributes of "blue" and "plastic" have both been applied to the same shape, meaning they are both applicable to the same toy.

The opposite is true as well. Assume the following:

- **The cone is blue.**
- **The cone is not made of plastic.**

Based on these clues, we can safely conclude the blue toy is not made of plastic.

Indirect statements can also bring about similar conclusions:

- **The cone is blue or red.**
- **The plastic toy is green or yellow.**

These clues allow us to deduce the cone must not be made of plastic.

Simple Order Constraints

Let's assume we have three possible events—A, B, and C—that could have happened at five possible times—1:00, 2:00, 3:00, 4:00, or 5:00.

A simple order constraint might look like the following statement:

- **A happened before B.**

From this, we can deduce A could not have happened at 5:00 (because then we wouldn't have a later time slot remaining when B could have occurred), and B could not have happened at 1:00 (because then we would not have an earlier time slot remaining when A could have occurred).

- **A happened before B and C.**

With this addition, we know A could not have happened at 4:00 or 5:00 (because we need to reserve two later time slots for both B and C), and B and C could not have happened at 1:00 (we don't know anything about these events, except that A happened before both of them).

- **A happened after B but before C.**

With this information, we can eliminate the 1:00 and 5:00 slots for A. We also know two events took place after B (since C happened after A, and A was already after B), so B could not have happened at 4:00 or 5:00. Further, two events took place before C (the logic applied here is exactly the same), so C could not have occurred at 1:00 or 2:00.

Order Span Conditions

Let's keep the same A, B, and C events that could have happened at 1:00, 2:00, 3:00, 4:00, or 5:00.

An order span condition might look like the following:

- **There was one hour between events B and C.**

Obviously, if we knew the time for B or C, this would allow us to solve for the time of the other event as well, but we can also use this clue for elimination if we know enough about either event.

Let's assume we've already eliminated 4:00 and 5:00 as options for B. In this case, scheduling C for 5:00 would be impossible, since B cannot occur at 4:00—so we can eliminate 5:00 for C.

- **There were 3 hours between B and C.**

Unlike with the previous example, a three-hour gap between events allows us to do some initial elimination. If either B or C took place at 3:00, we would not have a time slot within our range for the other event, as 3 hours from 3:00 would be either 12:00 or 6:00. So we can eliminate 3:00 as a time slot for both B and C.

Present Deliveries

Lisa bought a bunch of Christmas presents for her relatives from various online stores, and five of them arrived today, each one from a different delivery service.

1. The GLS courier made their delivery before the board game and the origami kit arrived.
2. Lisa signed off for the concert ticket before the present for her cousin was delivered.
3. There was about an hour between the deliveries by FedEx and UPS; an hour between the arrival of the origami kit parcel and the present Lisa plans to give to her best friend; and an hour between the arrival of the GLS courier and the present Lisa bought for her dad.
4. There were about two hours between the arrival of the present Lisa ordered for her best friend and the one she'll give to her dad; two hours between when the present for her mom arrived and when UPS delivered their package; and two hours between the arrival of the coloring books and the concert ticket.
5. The coloring books were delivered before UPS showed up, but after DHL delivered their parcel.
6. The present Lisa wants to give to her mom was delivered before the crochet yarn pack.

If you want to try this puzzle on your own, use the grid on the next page! The solution is available on page 23.

Present Deliveries

		Hour					Delivery					Person				
	Present Deliveries	6:00	5:00	4:00	3:00	2:00	GLS	FedEx	DHL	UPS	Local post	Dad	Mom	Cousin	Best friend	Brother
Present	Coloring books															
	Crochet yarn															
	Origami kit															
	Board game															
	Concert ticket															
Person	Brother															
	Best friend															
	Cousin															
	Mom															
	Dad															
Delivery	Local post															
	UPS															
	DHL															
	FedEx															
	GLS															

Present	Hour	Delivery	Person

15

First off, let's check for information we already know, based on the clues.

Clue 2 tells us the following:

- **Lisa signed off for the concert ticket before the present for her cousin was delivered.**

This means the concert ticket was not the last delivery of the day, and that the present for Lisa's cousin was not delivered first. We know this because there was at least one delivery after the ticket (the present for Lisa's cousin), and at least one delivery before the present for Lisa's cousin (the concert ticket).

The same reasoning can be applied to Clue 6:

- **The present Lisa wants to give to her mom was delivered before the crochet yarn pack.**

So there was at least one delivery after the present for Lisa's mom (the crochet yarn pack), and at least one delivery before the crochet yarn pack (the present for Lisa's mom).

Clue 1 comes with two pieces of info:

- **The GLS courier made their delivery before the board game and the origami kit arrived.**

What's important here is that the board game and the origami kit belong to the same category, and could not have arrived at the same hour. This means there were two deliveries made after the GLS delivery, so the GLS courier could not have shown up at 5:00 or at 6:00.

Clue 5 also has two pieces of information:

- **The coloring books were delivered before UPS showed up, but after DHL delivered their parcel.**

Obviously, the coloring books could not have been delivered by either UPS or DHL, but more importantly, this clue provides us with three timing conditions:

1. The coloring books were not the first or last delivery of the day.
2. There were two deliveries after DHL (the coloring books and the UPS delivery).
3. There were two deliveries before UPS (the coloring books and the DHL delivery).

If we look at the grid now, we have eliminated all possible deliveries at 2:00 except the concert ticket, so we can mark that solution and cross off all other possible delivery times for that present:

	Present Deliveries	6:00	5:00	4:00	3:00	2:00	GLS	FedEx	DHL	UPS	Local post	Dad	Mom	Cousin	Best friend	Brother
Present	Coloring books	X				X			X	X						
	Crochet yarn					X							X			
	Origami kit					X	X									
	Board game					X	X									
	Concert ticket	X	X	X	X	✓									X	
Person	Brother															
	Best friend															
	Cousin					X										
	Mom	X														
	Dad															
Delivery	Local post															
	UPS				X	X										
	DHL		X	X												
	FedEx															
	GLS		X	X												

Now let's take a closer look at Clue 3:

- **There was about an hour between the deliveries by FedEx and UPS; an hour between the arrival of the origami kit parcel and the present Lisa plans to give to her best friend; and an hour between the arrival of the GLS courier and the present Lisa bought for her dad.**

As we already know, the UPS delivery could not have happened at 2:00 or 3:00. This means the FedEx delivery could not have been delivered at 2:00 either, because the only time slot within one hour for the UPS delivery would be at 3:00, and we've already eliminated that option.

The same logic can be applied to the GLS courier and the present bought for Lisa's dad (additionally, we know the present for Lisa's dad was not brought by GLS).

The middle bit of info doesn't offer any extra time constraints, but we now know Lisa isn't planning to give an origami kit to her best friend.

Clue 4 reads as follows:

- **There were about two hours between the arrival of the present Lisa ordered for her best friend and the one she'll give to her dad; two hours between when the present for her mom arrived and when UPS delivered their package; and two hours between the arrival of the coloring books and the concert ticket.**

The second statement in this clue tells us UPS did not bring the present intended for Lisa's mother.

Since we know when the concert ticket delivery happened, we can mark off when the coloring books arrived as well:

		Hour					Delivery					Person				
	Present Deliveries	6:00	5:00	4:00	3:00	2:00	GLS	FedEx	DHL	UPS	Local post	Dad	Mom	Cousin	Best friend	Brother
Present	Coloring books	X	X	✓	X	X			X	X						
	Crochet yarn			X		X							X			
	Origami kit			X		X	X									
	Board game			X		X	X									
	Concert ticket	X	X	X	X	✓									X	
Person	Brother															
	Best friend															
	Cousin				X											
	Mom	X								X						
	Dad	X					X									
Delivery	Local post															
	UPS			X	X	X										
	DHL	X	X	X												
	FedEx					X										
	GLS	X	X													

Recall that Clue 5 gave us more information about the coloring books:

- **The coloring books were delivered before UPS showed up, but after DHL delivered their parcel.**

Since we now know the coloring book delivery happened at 4:00, we can cross out 4:00 as a possible delivery slot for both DHL and UPS.

We can now look at the two other pieces of information we have about the UPS delivery, from Clues 3 and 4:

- **There was about an hour between the FedEx and UPS deliveries.**
- **There were two hours between when the present for Lisa's mom arrived and the UPS delivery.**

If the FedEx delivery happened at 3:00, the UPS delivery would need to happen at either 2:00 or 4:00, and those slots are already occupied.

If we apply the same logic to the present for Lisa's mom, we can deduce it could only have arrived at 3:00 (if the UPS came at 5:00) or 4:00 (if the UPS came at 6:00).

We also know the following, from Clue 6:

- **The present Lisa wants to give to her mom was delivered before the crochet yarn pack.**

This means we need to eliminate the crochet yarn pack from the 3:00 slot (since the earliest Lisa's mom's present could be delivered is 3:00). This leaves two possible options for the 3:00 delivery: the origami kit and the board game.

Now we need to recall the information in Clue 1:

- **The GLS courier made their delivery before the board game and the origami kit arrived.**

This tells us the GLS courier was there before 3:00, leaving 2:00 as the only possible time slot for the GLS delivery.

With this knowledge, we can mark off this chain as well:

- There was about an hour between the arrival of the GLS courier and the present Lisa bought for her dad.
- There were about two hours between the arrival of the present Lisa ordered for her best friend and the one she'll give to her dad.

3:00 would be the only possible slot within an hour from 2:00, and since we don't have any deliveries earlier than 2:00, the only slot two hours away from 3:00 would be 5:00.

After marking off these findings, you'll likely notice you've got only one possible recipient for the 2:00 gift: Lisa's brother. There's also only one possible time the present for her mother could have been delivered: 4:00.

		Hour					Delivery					Person				
		6:00	5:00	4:00	3:00	2:00	GLS	FedEx	DHL	UPS	Local post	Dad	Mom	Cousin	Best friend	Brother
Present	Coloring books	X	X	✓	X	X			X	X						
	Crochet yarn			X	X	X								X		
	Origami kit			X		X					X					
	Board game			X		X										
	Concert ticket	X	X	X	X	✓									X	
Person	Brother	X	X	X	X	✓										
	Best friend	X	✓	X	X	X										
	Cousin		X	X	X	X										
	Mom	X	X	✓	X	X				X						
	Dad	X	X	X	✓	X	X									
Delivery	Local post					X										
	UPS			X	X	X										
	DHL	X	X	X		X										
	FedEx				X	X										
	GLS	X	X	X	X	✓										

21

With that, we've completed the person/hour subgrid!

Now we can apply the information related to the UPS delivery from Clues 3 and 4:

- **There were two hours between when the present for her mom arrived and the UPS delivered their package.**
- **There was about an hour between the deliveries by FedEx and UPS.**

And, in case you didn't mark it off already, resolving the GLS delivery time left us with only one possible time slot for the DHL courier, resolving that subgrid as well.

		Hour					Delivery					Person				
	Present Deliveries	6:00	5:00	4:00	3:00	2:00	GLS	FedEx	DHL	UPS	Local post	Dad	Mom	Cousin	Best friend	Brother
Present	Coloring books	X	X	✓	X	X			X	X						
	Crochet yarn			X	X	X						X				
	Origami kit			X		X	X									
	Board game			X		X	X									
	Concert ticket	X	X	X	X	✓								X		
Person	Brother	X	X	X	X	✓										
	Best friend	X	✓	X	X	X										
	Cousin	✓	X	X	X	X										
	Mom	X	X	✓	X	X					X					
	Dad	X	X	X	✓	X	X									
Delivery	Local post	X	X	✓	X	X										
	UPS	✓	X	X	X	X										
	DHL	X	X	X	✓	X										
	FedEx	X	✓	X	X	X										
	GLS	X	X	X	X	✓										

The only thing left to establish is the present/hour subgrid. From Clue 3, we know the following:

- **There was about an hour between the arrival of the origami kit parcel and the present Lisa plans to give to her best friend.**

The origami kit must have arrived at 6:00 (since Lisa's best friend's present arrived at 5:00 and the coloring books arrived at 4:00), leaving only the board game available to fill the 3:00 time slot.

And with that, the puzzle is complete!

Present Deliveries	Hour					Delivery					Person				
	6:00	5:00	4:00	3:00	2:00	GLS	FedEx	DHL	UPS	Local post	Dad	Mom	Cousin	Best friend	Brother
Coloring books	X	X	✓	X	X			X	X						
Crochet yarn	X	✓	X	X	X						X				
Origami kit	✓	X	X	X	X	X									
Board game	X	X	X	✓	X	X									
Concert ticket	X	X	X	X	✓								X		
Brother	X	X	X	X	✓										
Best friend	X	✓	X	X	X										
Cousin	✓	X	X	X	X										
Mom	X	X	✓	X	X				X						
Dad	X	X	X	✓	X	X									
Local post	X	X	✓	X	X										
UPS	✓	X	X	X	X										
DHL	X	X	X	✓	X										
FedEx	X	✓	X	X	X										
GLS	X	X	X	X	✓										

All Spots Taken

Once again, let's keep the same A, B, and C events that could have happened at 1:00, 2:00, 3:00, 4:00, or 5:00.

Let's also take a look at the following conditions:

- **A happened at 1:00 or 2:00.**
- **B happened at 1:00 or 2:00.**

Theoretically, these conditions don't give much information, as we cannot narrow down the times for A and B any further from this point; however, what we actually have here is two events that take up two possible time slots. This means, even if these conditions don't mention C, we can use them to conclude that C did not happen at 1:00 or 2:00, simply because we could not place both A and B if C occurred at either of these times.

Distinguishing Between Categories

It's very important to recognize the difference between elements in the same category, as well as differences between elements in different categories.

Let's return to the same A, B, and C events that could have happened at 1:00, 2:00, 3:00, 4:00, or 5:00. This time, we'll add a third element: Let's assume that one of those events happened at the school, another at the park, and another at the grocery store.

Let's look at the following sets of conditions.
Set 1:

- **There was one hour between A and B.**
- **There was one hour between B and C.**

Set 2:

- **There was one hour between A and B.**
- **There was one hour between event B and the school event.**

These sets look very much alike, but reveal different information. For the first set, we can be sure that A and C happened two hours from one another, and B was held in the hour between them. For the second set, we have two possible options:

1. There were 2 hours between A and the school event, and B happened between them.
2. A was the school event.

Also note: The information provided in the two conditions above is not enough to establish which of the options from Set 2 is true.

Cross-Referencing Categories

Let's keep the same A, B, and C events that could have happened at 1:00, 2:00, 3:00, 4:00, or 5:00, and may have happened at the school, at the park, or at the grocery store.

Sometimes information can be cross-referenced between different subgrids. For example, consider these conditions:

- **A and B took place before 2:00.**
- **The park event took place after 2:00.**

From this, we can conclude that neither A nor B took place at the park.

The Only Possible Spot

Let's keep the same A, B, and C events that could have happened at 1:00, 2:00, 3:00, 4:00, or 5:00.

Sometimes conditions are listed in a way that allows precise placement of one item, even without having complete information about the remaining items:

- **A took place at 1:00 or 3:00.**
- **C took place at 1:00 or 3:00.**
- **B took place between events A and C.**

If A took place at 1:00, it would mean C took place at 3:00. That would mean B, which is between them, must have taken place at 2:00. The other possibility is that C took place at 1:00, which would put A at 3:00—but even in this case, B would still take place at 2:00. So, regardless of A and C's final placement, we can lock in B at 2:00 based on this set of conditions.

The Talent Contest

The international talent contest has just concluded. Let's take a look at the results.

1. The participant from Australia placed better than the contestant from the USA, but worse than Vivian.
2. Jessie placed between Cole and Jack.
3. The participant from Japan placed better than Anna.
4. One person placed between the acrobat and Cole.
5. Two people placed between the Australian participant and the dancer.
6. The American participant and the dancer placed next to one another, in no specific order.
7. Anna placed beneath the ventriloquist.
8. One person placed between Anna and the singer, and one person placed between Jack and the German contestant.

If you want to try this puzzle on your own, use the grid on the next page! The solution is available on page 35.

The Talent Contest

	The Talent Contest	Place					Country					Performance				
		5	4	3	2	1	Japan	China	USA	Australia	Germany	Acrobat tricks	Ventriloquist	Card tricks	Song	Dance routine
Person	Anna															
	Cole															
	Vivian															
	Jack															
	Jessie															
Performance	Dance routine															
	Song															
	Card tricks															
	Ventriloquist															
	Acrobat tricks															
Country	Germany															
	Australia															
	USA															
	China															
	Japan															

Person	Place	Country	Performance

Let's start by marking off the initial conditions. Try filling in the X's on the chart yourself, then compare with the grid below. If you missed something, you can always go back to the clues and double-check:

The Talent Contest		Place					Country					Performance				
		5	4	3	2	1	Japan	China	USA	Australia	Germany	Acrobat tricks	Ventriloquist	Card tricks	Song	Dance routine
Person	Anna				X	X						X		X		
	Cole											X				
	Vivian	X	X						X	X						
	Jack										X					
	Jessie	X				X										
Performance	Dance routine			X							X					
	Song									X						
	Card tricks															
	Ventriloquist	X														
	Acrobat tricks															
Country	Germany															
	Australia	X		X		X										
	USA				X	X										
	China															
	Japan		X													

When rereading the conditions after our first passthrough, notice that we can further limit the placement of the dancer, based on Clues 5 and 6:

- **Two people placed between the Australian participant and the dancer.**
- **The American participant and the dancer placed next to one another, in no specific order.**

Only one spot meets both of these conditions: fifth place.

30

It's also worth noting that Jack, Jessie, and Cole took consecutive spots, with Jessie in the middle (though we don't yet know whether Cole or Jack placed higher). This is based on Clue 2:

- **Jessie was the only participant who placed between Cole and Jack.**

An additional piece of information we can pull from this clue is that, no matter how the trio placed, one of them (either Cole or Jack) will always end up in third place. This is true whether they placed 1, 2, 3; 2, 3, 4; or 3, 4, 5. This also tells us neither Anna nor Vivian could have placed third.

The Talent Contest		Place					Country					Performance				
		5	4	3	2	1	Japan	China	USA	Australia	Germany	Acrobat tricks	Ventriloquist	Card tricks	Song	Dance routine
Person	Anna			X		X	X						X	X		
	Cole											X				
	Vivian	X	X	X						X	X					
	Jack										X					
	Jessie	X			X											
Performance	Dance routine	X	X	X	X	X					X					
	Song	X			X			X								
	Card tricks	X														
	Ventriloquist	X														
	Acrobat tricks	X														
Country	Germany															
	Australia	X		X		X										
	USA				X	X										
	China															
	Japan	X														

Considering the information we have on the Australian and USA contestants from Clues 5 and 6, we can now locate the placement each one took:

- **Two people placed between the Australian participant and the dancer.**
- **The American participant and the dancer placed next to one another, in no specific order.**

So we now know the Australian participant placed second, and the American participant placed fourth.

This means, if Jessie, Jack, and Cole took placements 1, 2, and 3 (regardless of order), it would leave no placement option for Vivian (since she must have placed above the Australian participant, as we know from Clue 1). So we can conclude that Vivian won the contest. Jessie, Jack and Cole still need three consecutive spots though, meaning Anna couldn't have finished fourth. Also, since Jessie needs to be between Jack and Cole, he can't be in the second place spot.

As we limit the possible positions for Anna, we can further apply the information from Clue 8:

- **One person placed between Anna and the singer.**

This eliminates the singer from spot 2. Here's how our grid looks now:

The Talent Contest	Place					Country					Performance				
	5	4	3	2	1	Japan	China	USA	Australia	Germany	Acrobat tricks	Ventriloquist	Card tricks	Song	Dance routine
Person Anna		X	X		X	X						X		X	
Cole				X							X				
Vivian	X	X	X	X	✓			X	X						
Jack				X						X					
Jessie	X			X											
Performance Dance routine	✓	X	X	X	X				X						
Song	X			X	X			X							
Card tricks	X														
Ventriloquist	X														
Acrobat tricks	X														
Country Germany		X		X											
Australia	X	X	X	✓	X										
USA	X	X	X	X	X										
China		X		X											
Japan	X	X		X											

Establishing positions for two country representatives limited the possible placements for the German contestant down to three options. But we know the following, from Clue 8:

- **One participant placed between Jack and the German contestant.**

This means Jack could only have placed third or fifth. Same for Cole, since he needs to be separated from Jack by one person. Speaking of that person, Jessie is now limited to fourth place, since no matter how Cole and Jack placed, fourth is the only spot left between the two of them.

This leaves Anna as our only option for second place, which, in turn, lets us place the singer, the ventriloquist and the Japanese representative using Clues 3, 7, and 8:

- **The participant from Japan placed better than Anna.**
- **Anna placed beneath the ventriloquist.**
- **One person placed between Anna and the singer.**

Also, we can limit possible placements for the acrobat using Clue 4:

- **One person placed between the acrobat and Cole.**

The Talent Contest		Place					Country					Performance				
		5	4	3	2	1	Japan	China	USA	Australia	Germany	Acrobat tricks	Ventriloquist	Card tricks	Song	Dance routine
Person	Anna	X	X	X	✓	X	X						X		X	
	Cole		X		X	X						X				
	Vivian	X	X	X	X	✓			X	X						
	Jack		X								X					
	Jessie	X	✓	X	X	X										
Performance	Dance routine	✓	X		X	X				X						
	Song	X	✓	X	X	X			X							
	Card tricks	X	X			X										
	Ventriloquist	X	X	X	X	✓										
	Acrobat tricks	X	X		X	X										
Country	Germany		X		X	X										
	Australia	X	X	X	✓	X										
	USA	X	✓	X	X	X										
	China		X		X											
	Japan		X	X	X	✓										

34

Since only one spot remains for the acrobat, we can effectively solve the second subgrid.

This, in turn, will let us place Cole, Jack, and the German contestant, using Clues 4 and 8:

- **One person placed between the acrobat and Cole.**
- **One person placed between Jack and the German contestant.**

And the puzzle is solved!

The Talent Contest		Place					Country					Performance				
		5	4	3	2	1	Japan	China	USA	Australia	Germany	Acrobat tricks	Ventriloquist	Card tricks	Song	Dance routine
Person	Anna	X	X	X	✓	X	X						X		X	
	Cole	✓	X	X	X	X						X				
	Vivian	X	X	X	✓				X	X						
	Jack	X	X	✓	X	X					X					
	Jessie	X	✓	X	X	X										
Performance	Dance routine	✓	X	X	X	X					X					
	Song	X	✓	X	X	X		X								
	Card tricks	X	X	X	✓	X										
	Ventriloquist	X	X	X	X	✓										
	Acrobat tricks	X	X	✓	X	X										
Country	Germany	✓	X	X	X	X										
	Australia	X	X	X	✓	X										
	USA	X	X	✓	X	X										
	China	X	X	✓	X	X										
	Japan	X	X	X	X	✓										

The Only Possible Distance

The above reasoning can sometimes also be applied when we know the distance between two elements.

Let's keep the same A, B, and C events that could have happened at 1:00, 2:00, 3:00, 4:00, or 5:00.

Additionally, we'll assume the following conditions:

- **A did not take place at 1:00.**
- **B took place three hours after A.**

If we assume A took place later than 2:00, B would have to take place later than 5:00, which is not possible. Further, since A could not have occurred at 1:00, we are left with only one solution for A and B that meets the distance condition: 2:00 for A and 5:00 for B.

This rule can be applied even if the exact position of one of the elements is unknown. Consider this example:

- **A took place at 1:00 or 5:00.**
- **There were 2 hours between A and B.**

Regardless of the actual time A took place, the only time slot within the two-hour range of the rules for both 1:00 and 5:00 is 3:00, so we can conclude B took place at 3:00.

Distance and Limitations

Sometimes the distance connections between several pairs of elements can be used to limit options for placement.

Let's return, once again, to the same A, B, and C events that could have happened at 1:00, 2:00, 3:00, 4:00, or 5:00.

We'll also add the following conditions:

- **B took place one hour after A.**
- **There was one hour between C and A.**

As we've previously discussed, the first of these conditions means A could not have happened at 5:00, because then we would have no time slots remaining for B.

Let's assume A took place at 1:00. This would mean both B and C took place at 2:00, because 2:00 is the only option that places C one hour away from A, and also B one hour later than 1:00. Hence, we can be certain A did not take place at 1:00.

The Coloring Hobby

Girls from the local coloring club brought their new coloring books to today's meeting.

1. The book purchased on Etsy is longer than the one bought on Lulu.
2. The flower book has more pages than the 30-page fashion-themed book, but less pages than the book Iqra brought.
3. The book bought on Lulu has 5 fewer pages than the animal-themed book.
4. There are 5 pages of difference between the book bought in the local store and the book Lin brought; 5 pages of difference between the mandala-themed book and the animal-themed book; and 5 pages of difference between Iqra's book and the book bought on Lulu.
5. There are 10 pages of difference between the book received as a present and Elisa's book.
6. The book bought on Amazon is longer than the book Prethepa brought.
7. The book Elisa brought has more pages than the flower-themed book.

If you want to try this puzzle on your own, use the grid on the next page! The solution is available on page 50.

The Coloring Hobby

The Coloring Hobby		Pages					Theme					Source				
		45	40	35	30	25	Landscape	Fashion	Animal	Mandala	Flower	Lulu	Local store	Amazon	Etsy	Present
Person	Elisa															
	Prethepa															
	Darla															
	Lin															
	Iqra															
Source	Present															
	Etsy															
	Amazon															
	Local store															
	Lulu															
Theme	Flower															
	Mandala															
	Animal															
	Fashion															
	Landscape															

Person	Pages	Theme	Source

Let's go over the initial conditions. First of all, one matched pair is already provided in Clue 2:

- **The flower coloring book has more pages than the 30-page fashion-themed book, but less pages than the book Iqra brought.**

From this, we can mark our grid to indicate the fashion-themed book has 30 pages, the flower-themed one has more than 30 pages (but not the maximum amount), and the book Iqra brought has more than 35 pages (as it has to have more pages than the flower-themed one). Additionally, Iqra didn't bring the flower-themed book or the fashion-themed book.

Clues 1 and 3 are fairly straightforward:

- **The book purchased on Etsy is longer than the one bought on Lulu.**
- **The book bought on Lulu has 5 fewer pages than the animal-themed book.**

We can tell the book from Lulu doesn't have the maximum number of pages, and neither the book bought on Etsy nor the animal-themed book is the shortest book. We also know the animal-themed book did not come from Lulu. These clues do not, however, give us any direct connections between the animal-themed book and Etsy.

The same method of deduction can be applied to Clues 6 and 7:

- **The book bought on Amazon is longer than the book Prethepa brought.**
- **The book Elisa brought has more pages than the flower-themed book.**

The book Prethepa brought is not the longest, the Amazon one is not the shortest, and Prethepa cannot have brought the book from Amazon. Elisa did not bring the flower-themed book and, since the latter must have over 30 pages, we now know Elisa's book must have either 40 or 45 pages.

The Coloring Hobby		Pages					Theme					Source				
		45	40	35	30	25	Landscape	Fashion	Animal	Mandala	Flower	Lulu	Local store	Amazon	Etsy	Present
Person	Elisa			X	X	X					X					
	Prethepa	X													X	
	Darla															
	Lin															
	Iqra			X	X	X	X				X					
Source	Present															
	Etsy					X										
	Amazon					X										
	Local store															
	Lulu	X								X						
Theme	Flower	X			X	X										
	Mandala				X											
	Animal				X	X										
	Fashion	X	X	X	✓	X										
	Landscape				X											

Now that we have limited the options for Elisa, we can use Clue 5:

- **There are 10 pages of difference between the book received as a present and Elisa's book.**

This tells us Eliza didn't bring the book received as a present, and that the present was not the shortest book, nor the one with 40 pages.

And finally we have Clue 4's set of 5-page difference conditions:

- **There are 5 pages of difference between the book bought in the local store and the book Lin brought; 5 pages of difference between the mandala-themed book and the animal-themed book; and 5 pages of difference between Iqra's book and the book bought on Lulu.**

The first condition only tells us the book Lin brought does not come from the local store. The second one allows us to conclude that the mandala-themed book could not have been the shortest (because the animal-themed book cannot have 30 pages), allowing us to determine that the shortest book was landscape-themed. The third informs us that the book bought on Lulu was not brought by Iqra, and has more than 30 pages (since Iqra's book can only be 40 or 45 pages long).

Here's how our grid looks at this point:

The Coloring Hobby		Pages					Theme					Source				
		45	40	35	30	25	Landscape	Fashion	Animal	Mandala	Flower	Lulu	Local store	Amazon	Etsy	Present
Person	Elisa			X	X	X					X					X
	Prethepa	X													X	
	Darla															
	Lin												X			
	Iqra			X	X	X			X		X	X				
Source	Present		X			X										
	Etsy					X										
	Amazon					X										
	Local store															
	Lulu	X			X	X			X							
Theme	Flower	X			X	X										
	Mandala				X	X										
	Animal				X	X										
	Fashion	X	X	X	✓	X										
	Landscape	X	X	X	X	✓										

As you can see, Elisa and Iqra both could only have brought the 40- or 45-page books. Additionally, neither of them brought the flower-themed book. This brings us to two conclusions:

1. The flower-themed book must have 35 pages (otherwise it would have been brought by Elisa or Iqra).
2. No other girl could have brought the two longest books.

We've also crossed out every option but one for the source of the 25-page book, meaning it must have come from the local store.

	Pages					Theme					Source				
The Coloring Hobby	45	40	35	30	25	Landscape	Fashion	Animal	Mandala	Flower	Lulu	Local store	Amazon	Etsy	Present
Person Elisa			X	X	X					X					X
Prethepa	X	X											X		
Darla	X	X													
Lin	X	X										X			
Iqra			X	X	X	X			X	X	X				
Source Present			X		X										
Etsy					X										
Amazon					X										
Local store	X	X	X	X	✓										
Lulu	X		X	X	X			X							
Theme Flower	X	X	✓	X	X										
Mandala			X	X	X										
Animal			X	X	X										
Fashion	X	X	X	✓	X										
Landscape	X	X	X	X	✓										

Recall Clue 5:

- **There are 10 pages of difference between the book received as a present and Elisa's book.**

This allows us to limit the present to 30 or 35 pages, based on the information we have now.

Additionally, recall this condition from Clue 4:

- **There are 5 pages of difference between the book bought in the local store and the book Lin brought.**

This means Lin must have brought the 30-page book.

The Coloring Hobby		Pages					Theme					Source				
		45	40	35	30	25	Landscape	Fashion	Animal	Mandala	Flower	Lulu	Local store	Amazon	Etsy	Present
Person	Elisa			X	X	X					X					X
	Prethepa	X	X											X		
	Darla	X	X													
	Lin	X	X	X	✓								X			
	Iqra			X	X	X					X	X				
Source	Present	X	X			X										
	Etsy	✓	X	X	X	X										
	Amazon	X														
	Local store	X	X	X	✓											
	Lulu	X			X				X							
Theme	Flower	X	X	✓												
	Mandala			X	X	X										
	Animal				X	X										
	Fashion	X	X	✓		X										
	Landscape	X	X	X	✓											

Now let's fill in the missing information in columns two and three.

We know the following, from Clues 3 and 4:

- **The book bought on Lulu has 5 fewer pages than the animal-themed book.**
- **There are 5 pages of difference between Iqra's book and the book bought on Lulu.**

Since no options remain where Iqra's book and the animal book could be different, Iqra must have brought the animal-themed book. This means Elisa must have brought the mandala-themed one, since animal and mandala are the only two remaining theme options for 40 and 45 pages.

The Coloring Hobby		Pages					Theme					Source				
		45	40	35	30	25	Landscape	Fashion	Animal	Mandala	Flower	Lulu	Local store	Amazon	Etsy	Present
Person	Elisa			X	X	X	X	X	X	✓	X					X
	Prethepa	X	X		X			X	X					X		
	Darla	X	X		X			X	X							
	Lin	X	X	✓	X	✓	X	✓	X					X		
	Iqra			X	X	✓	X	X	✓	X	X	X				
Source	Present	X	X													
	Etsy	✓	X	X	X	X										
	Amazon	X			X	X										
	Local store	X	X	X	✓	✓		X	X	X	X					
	Lulu	X			X	X	X		X							
Theme	Flower	X	X	✓	X	X										
	Mandala			X	X	X										
	Animal			X	X											
	Fashion	X	X	X	✓	X										
	Landscape			X	X	✓										

Looking at the updated grid, we now know neither Elisa nor Iqra could have brought a book from the local store or received a book as a present. Also, one of them must have brought the book from Etsy because that book is 45 pages long.

We know Lin couldn't have brought the flower-themed book, since the 30-page book is fashion themed, so we can cross that out as well.

The book received as a present must be flower- or fashion-themed (otherwise we'd get a contradiction with number of pages); the book from Etsy can only have the animal or mandala theme; and, finally, the book from Lulu must be flower- or mandala-themed, and wasn't brought by Lin (who, as we just deduced, has the fashion-themed book).

The Coloring Hobby	Pages					Theme					Source				
	45	40	35	30	25	Landscape	Fashion	Animal	Mandala	Flower	Lulu	Local store	Amazon	Etsy	Present
Person Elisa			X	X	X	X	X	X	✓	X		X			X
Prethepa	X	X		X			X		X				X	X	
Darla	X	X		X										X	
Lin	X	X	✓	X	X	X	✓		X	X	✓	X		X	
Iqra			X	X	X	X	X	✓	X	X	X	X			X
Source Present	X	X		X		X		X	X						
Etsy	✓	X	X	X	X	X	X			X					
Amazon	X			X		X									
Local store	X	X	X	X		✓	X	X	X	X					
Lulu	X			X	X	X	X	X							
Theme Flower	X	X	✓		X										
Mandala			X	X	X										
Animal			X		X										
Fashion	X	X	X	✓	X										
Landscape	X	X	X	X	✓										

There are 5 pages of difference between the book from Lulu and Iqra's book (Clue 4), and 10 pages between the book received as a present and Elisa's book (Clue 5). Additionally, the present will always have fewer pages than the book from Lulu, as we can see on the grid.

We can use math to prove that there cannot be only 5 pages of difference between the book from Lulu and the present, by introducing variables in place of book lengths and finding a logical impossibility through substitution. Let's walk through this reasoning together.

We know Iqra brought the animal-themed book. We also know that the book bought on Lulu has 5 fewer pages than the animal-themed book, based on Clue 4. Thus, we know that the book bought on Lulu is 5 fewer pages than Iqra's book, so Iqra's book is X+5 pages long, where X is the length of the book from Lulu.

Clue 5 tells us there are 10 pages of difference between Elisa's book and the present, and the grid shows us there's no way the present could have more pages than Elisa's book, meaning that the present must be 10 pages shorter than Elisa's book. That means Elisa's book has a length of Y+10, where Y is the length of the book received as a present.

Finally, we know that the present must have fewer pages than the book from Lulu, as seen on the grid. That means X>Y, where X is the length of the book from Lulu and Y is the length of the book received as a present. We can now omit any answer where X=Y, since we've just shown that they aren't equal.

Since we know that X>Y we can designate the book from Lulu as Y+Z, where Y is the length of the book received as a present and Z is the difference in pages between the present and the book from Lulu. Through substitution, this indicates that (Y+Z)+5 is the length of Iqra's book. Since we already know that Elisa's book is Y+10 pages long, and we know that Elisa and Iqra cannot bring books of the same length, we can omit the value 5 for Z (since Y+5+5=Y+10). Ergo, the difference in pages between the present and the book from Lulu is not 5 pages. That means the difference between the book from Lulu and the present must be 10 pages, since it's the only option remaining on the grid.

With this information, we can solve the second subgrid:

The Coloring Hobby		Pages					Theme					Source				
		45	40	35	30	25	Landscape	Fashion	Animal	Mandala	Flower	Lulu	Local store	Amazon	Etsy	Present
Person	Elisa			X	X	X	X	X	X	✓	X		X			X
	Prethepa	X	X		X			X		X					X	X
	Darla	X	X		X			X						X		
	Lin	X	X	X	X	X	✓	X	X		X	X	X		X	
	Iqra		X	X	X	X			✓	X	X	X	X			X
Source	Present	X	X		✓											
	Etsy	✓	X	X	X	X	X	X			X					
	Amazon	X	✓	X												
	Local store	X	X	X	X	✓	✓	X	X	X	X					
	Lulu	X	✓	X	X	X		X	X							
Theme	Flower	X	X	✓												
	Mandala			X	X	X										
	Animal			X	X	X										
	Fashion	X	X	X	✓	X										
	Landscape		X	X	X	✓										

From here we can define the lengths of Elisa and Iqra's books (and also the mandala- and animal-themed books, since we know Elisa had mandalas and Iqra had animals).

At this point, the only information missing is the book length for Prethepa and Darla; however, we know the length of the Amazon purchase, which means we can define that also! Darla must have bought from Amazon, since Prethepa could not have sourced her book from there, so Darla's book length is the length of the Amazon purchase.

And the puzzle is solved!

49

	The Coloring Hobby	Pages					Theme					Source				
		45	40	35	30	25	Landscape	Fashion	Animal	Mandala	Flower	Lulu	Local store	Amazon	Etsy	Present
Person	Elisa	X	✓	X	X	X	X	X	X	✓	X		X			X
	Prethepa	X	X	X	✓			X	X					X	X	
	Darla	X	X	✓	X	X		X	X	X				X		
	Lin	X	X	X	✓	X		X				X	X		X	
	Iqra	✓	X	X	X	X	X	X	X	✓	X	X	X			X
Source	Present	X	X	X	✓	X		X		X	X					
	Etsy	✓	X	X	X	X	X	X			X					
	Amazon	X	X	✓	X	X										
	Local store	X	X	X	X	✓	✓	X	X	X	X					
	Lulu	X	✓	X	X	X	X	X	X							
Theme	Flower	X	X	✓	X	X										
	Mandala	X	✓	X	X	X										
	Animal	✓	X	X	X	X										
	Fashion	X	X	X	✓	X										
	Landscape	X	X	X	X	✓										

50

Chaining Up

Let's keep the same A, B, and C events that could have happened at 1:00, 2:00, 3:00, 4:00, or 5:00.

We'll also add the following conditions:

- **B took place an hour after A.**
- **A took place at 3:00 or later.**

Considering B took place after A, only two possible times remain for A: 3:00 and 4:00. These would place B at 4:00 or 5:00 respectively.

We might not know which of these two options is correct, but we do know one of the events—A or B—always ends up at 4:00. This means C cannot occur at 4:00.

Let Us Assume...

When solving harder puzzles, at some point you arrive at a situation when no clues can be applied directly to further narrow the choices. Don't worry! This doesn't mean you're stuck for good and the puzzle is broken. The way forward can be found by making some assumptions and seeing if those assumptions bring you to a contradiction.

Let's assume we have a pile of wooden tokens. The tokens are three shapes—cubes, cylinders, and pyramids. They are green, red, and blue in color, and there are three quantities of token types—1, 2, and 3.

Now let's take it one step further and add the following conditions:

- **The pile has fewer cubes than pyramids.**
- **The pile has fewer red tokens than pyramids.**
- **The cubes are not red.**

There are fewer cubes than pyramids, as seen in the first clue, so we know the pyramid count can't be 1. Let's assume the pyramid count is 2. Then both the cube and red token count would be 1, but we know that's impossible because of the third clue, which tells us the cubes are not red. Thus, we can conclude there are 3 pyramid tokens.

This was a very simplistic example, and the contradiction was quite obvious. The sample puzzle below contains more complex examples of deductions based on assumptions.

Picking a Good Assumption

Complex grids that require assumptions are usually at least 3x3 in size, sometimes even bigger. Trying to build an assumption from a random point in a grid of that size can lead to a lot of time spent tracking possibilities that might not result in contradictions.

To narrow down your options, use these tips to determine the best picks for making assumptions for logic grid solutions:

- Choose options that have three or more clues based off their placement on the grid.
- Choose options that have only two possible solutions left.
- Choose options from a subgrid with two or more clues that only contain values from that subgrid.

If you have an option meeting more than one of these conditions, all the better!

The Horse Race

A group of friends attended a horse race and placed bets on the outcome.

1. Star beat both Vasili and Vijay's picks.
2. The bay and the black horses finished one after the other, in no specific order.
3. The horse Liz bet on beat the chestnut but lost to the bay.
4. Prince beat the palomino horse but lost to Sunshine.
5. The horse Vasili bet on and the bay finished consecutively, in no specific order.
6. Two horses finished between Divinity and the horse Vijay bet on, in no specific order.
7. Joana's pick beat Jack's pick.
8. Sunshine lost to the bay horse.

If you want to try this puzzle on your own, use the grid on the next page. The solution is available on page 61.

The Horse Race

A Horse Race	Place					Horse					Color				
	5	4	3	2	1	Divinity	Sunshine	Comet	Prince	Star	Palomino	White	Black	Chestnut	Bay
Person Vasili															
Jack															
Joana															
Vijay															
Liz															
Color Bay															
Chestnut															
Black															
White															
Palomino															
Horse Star															
Prince															
Comet															
Sunshine															
Divinity															

Person	Place	Horse	Color

Let's start by marking off the initial conditions. Try filling in the X's on the chart yourself, then compare with the grid below. If you missed something compared to our result below, you can always go back to the clues and double-check:

A Horse Race		Place					Horse					Color				
		5	4	3	2	1	Divinity	Sunshine	Comet	Prince	Star	Palomino	White	Black	Chestnut	Bay
Person	Vasili	X			X						X					X
	Jack				X											
	Joana	X														
	Vijay			X	X	X	X				X					
	Liz	X			X										X	X
Color	Bay	X	X													
	Chestnut				X	X										
	Black	X														
	White															
	Palomino				X	X			X		X					
Horse	Star															
	Prince					X										
	Comet															
	Sunshine	X	X			X										
	Divinity		X	X												

The initial clues allow us to conclude Joana picked the winner of the race. Additionally, we know Sunshine lost to the bay horse and the bay could not have finished third. Knowing this eliminates Vasili's pick and the black horse from taking fourth. And, as Sunshine could not have won, we can eliminate Prince from second place and the palomino horse from third.

A Horse Race	Place					Horse					Color				
	5	4	3	2	1	Divinity	Sunshine	Comet	Prince	Star	Palomino	White	Black	Chestnut	Bay
Person Vasili	X	X			X					X					X
Jack					X										
Joana	X	X	X	X	✓										
Vijay			X		X	X				X					
Liz	X				X									X	X
Color Bay	X	X	X							X					
Chestnut				X	X										
Black	X	X													
White															
Palomino			X	X	X	X			X						
Horse Star	X	X	X												
Prince	X			X	X										
Comet															
Sunshine	X	X			X										
Divinity		X	X												

At this point there isn't much conclusive information we can add to the picture using the clues directly. Let's make an assumption instead: we'll assume Sunshine finished second.

If Sunshine finished second it would mean Star won and Divinity was last because we have no other options. Hence, Vijay's horse must have come second. As the bay horse beat Sunshine (per Clue 8), it also means the bay took first place—but now we arrive at a contradiction: Vasili's pick should finish next to the bay horse, and the only possible spot where that could be true is taken by Vijay's pick, under the conditions of this assumption.

A Horse Race		Place					Horse					Color				
		5	4	3	2	1	Divinity	Sunshine	Comet	Prince	Star	Palomino	White	Black	Chestnut	Bay
Person	Vasili	X	X	●	X						X					X
	Jack				X											
	Joana	X	X	X	X	✓										
	Vijay			X	●	X	X				X					
	Liz	X				X									X	X
Color	Bay	X	X	X	●		X									
	Chestnut				X	X										
	Black	X	X													
	White															
	Palomino			X	X	X		X		X						
Horse	Star	X	X	X	●											
	Prince	X			X	X										
	Comet															
	Sunshine	X	X	**?**	X											
	Divinity	●	X	X												

The above contradiction confirms that there's no way Sunshine could have finished second, so it must have finished third. This leaves fourth place as the only possible option for Prince.

Prince's placement allows us to conclude the palomino horse took last (per Clue 4), meaning the chestnut was not in fifth, so Liz's pick must have finished above fourth place.

After this elimination is made, we have two people left whose chosen horses could have taken fourth and fifth: Jack and Vijay. This means we can cross out Jack and Vijay from all other possible slots, which, in turn, allows us to eliminate Divinity from the last place (per Clue 6), leaving Comet as the only horse who could have placed fifth.

A Horse Race		Place					Horse					Color				
		5	4	3	2	1	Divinity	Sunshine	Comet	Prince	Star	Palomino	White	Black	Chestnut	Bay
Person	Vasili	X	X			X					X					X
	Jack			X	X	X										
	Joana	X	X	X	X	✓										
	Vijay			X	X	X	X				X					
	Liz	X	X			X									X	X
Color	Bay	X	X	X					X							
	Chestnut	X			X	X										
	Black	X	X													
	White	X														
	Palomino	✓	X	X	X	X	X				X					
Horse	Star	X	X	X												
	Prince	X	✓	X	X	X										
	Comet	✓	X	X	X	X										
	Sunshine	X	X	✓	X	X										
	Divinity	X	X	X												

Next we will assume the bay horse finished second. This brings us to a very obvious contradiction, as it places both Vasili and Liz's picks in third.

A Horse Race		Place					Horse					Color				
		5	4	3	2	1	Divinity	Sunshine	Comet	Prince	Star	Palomino	White	Black	Chestnut	Bay
Person	Vasili	X	X	●		X					X					X
	Jack			X	X	X										
	Joana	X	X	X	X	✓										
	Vijay			X	X	X	X				X					
	Liz	X	X	●		X									X	X
Color	Bay	X	X	X	?		X									
	Chestnut	X			X	X										
	Black	X	X													
	White	X														
	Palomino	✓	X	X	X	X		X		X						
Horse	Star	X	X	X												
	Prince	X	✓	X	X	X										
	Comet	✓	X	X	X	X										
	Sunshine	X	X	✓	X	X										
	Divinity	X	X	X												

The bay must have won the race if it did not finish second, which instantly gives us the positions for Liz and Vasili's picks (per Clues 3 and 5), as well as the placement of the black horse (per Clue 2).

Also, since Liz's pick beat the chestnut (per Clue 3), we now have enough information to solve the second subgrid.

The horse picked by Vasili finished second, meaning Star won the race and Divinity was second, allowing us to pinpoint Vijay's bet (per Clue 6), which then determines Jack's.

And with this, the puzzle is solved!

A Horse Race

		Place					Horse					Color				
		5	4	3	2	1	Divinity	Sunshine	Comet	Prince	Star	Palomino	White	Black	Chestnut	Bay
Person	Vasili	X	X	X	✓	X					X					X
	Jack	X	X	✓	X	X										
	Joana	X	X	X	✓											
	Vijay	✓	X	X	X	X	X				X					
	Liz	X	X	✓	X	X									X	X
Color	Bay	X	X	X	✓		X									
	Chestnut	✓	X	X	X	X										
	Black	X	X	✓	X	X										
	White	X	✓	X	X	X										
	Palomino	✓	X	X	X	X	X		X							
Horse	Star	X	X	X	✓	X										
	Prince	X	✓	X	X	X										
	Comet	✓	X	X	X	X										
	Sunshine	X	X	✓	X	X										
	Divinity	X	X	✓	X	X										

61

Mavis Goes to the Movies

Mavis is a huge film fan, and she likes to go to the movies every month. Currently, she's eagerly awaiting the premiere of several movies, one per month, August through December. Each of these upcoming movies is a different genre, and Mavis will enjoy each premiere accompanied by one of her friends.

1. One film will have its premiere between the horror and the western, in some order.
2. The musical and the rom-com will premiere in consecutive months, in some order.
3. One movie will premiere between the musical and the movie Mavis will watch with Brittany, in some order.
4. The rom-com will premiere after the movie Mavis wants to watch with Brittany.
5. Molly will enjoy a movie with Mavis sometime after the musical's premiere.
6. "Charlie & Margot" premieres the month before Mavis is scheduled to go to the movies with Molly.
7. Two movies have premieres between "A Summer to Remember" and the movie Mavis will enjoy with Claire, in some order.
8. "Our Trip to the Frog Pond" will premiere sometime after the horror movie. Either the month before or the month after "Our Trip to the Frog Pond" premieres, Mavis will see a movie with Sandra.
9. "Meet Me at Dusk" will not premiere in October.
10. Mavis will enjoy a movie with Tiffany before the premiere of "A Talk Under the Moon."

If you want to try this puzzle on your own, use the grid on the next page. The solution is available on page 72.

Mavis Goes to the Movies

Mavis Goes to the Movies		Month					Friend					Genre				
		December	November	October	September	August	Brittany	Sandra	Claire	Molly	Tiffany	Action	Western	Horror	Musical	Rom-com
Movie	Our trip...															
	A summer....															
	A talk...															
	Meet me...															
	Charlie & Margot															
Genre	Rom-com															
	Musical															
	Horror															
	Western															
	Action															
Friend	Tiffany															
	Molly															
	Claire															
	Sandra															
	Brittany															

Movie	Month	Genre	Friend

As usual, let's start by marking off the initial conditions. If you missed something compared to our result below, you can always go back to the clues and double-check:

Mavis Goes to the Movies	Month					Friend					Genre				
	December	November	October	September	August	Brittany	Sandra	Claire	Molly	Tiffany	Action	Western	Horror	Musical	Rom-com
Movie Our trip...					X										
A summer....			X												
A talk...					X										
Meet me...			X												
Charlie & Margot	X														
Genre Rom-com					X										
Musical	X														
Horror	X														
Western															
Action															
Friend Tiffany	X														
Molly					X										
Claire			X												
Sandra															
Brittany	X														

For all previous puzzles we could use the clues to add some additional information to the grid, but that doesn't seem to be the case for this one. Does that mean the puzzle's broken? Luckily, no! Let's take a closer look at the second vertical subgrid.

Let's assume the musical premiere happened in October. Clue 1 tells us the following:

- **One film will have its premiere between the horror and the western, in some order.**

This would mean the horror and western would have to premiere in September and November, but we also know the following (from Clue 2):

- **The musical and the rom-com will premiere in consecutive months, in some order**

That means an October premiere for the musical would be impossible.

	Mavis Goes to the Movies	Month					Friend					Genre				
		December	November	October	September	August	Brittany	Sandra	Claire	Molly	Tiffany	Action	Western	Horror	Musical	Rom-com
Movie	Our trip...					X										
	A summer....			X												
	A talk...					X										
	Meet me...			X												
	Charlie & Margot	X														
Genre	Rom-com					X										
	Musical	X	?													
	Horror	X	●		●											
	Western		●		●											
	Action															
Friend	Tiffany	X														
	Molly					X										
	Claire			X												
	Sandra															
	Brittany	X														

The rom-com is bound by the exact same limitations, so we can eliminate an October premiere for that genre as well.

What's more, the action movie also could not have aired in October, because that would place the rom-com and the musical in December and August, which are not consecutive months, thus violating the condition in Clue 2.

Mavis Goes to the Movies		December	November	October	September	August	Brittany	Sandra	Claire	Molly	Tiffany	Action	Western	Horror	Musical	Rom-com
		Month					Friend					Genre				
Movie	Our trip...					X										
	A summer....			X												
	A talk...					X										
	Meet me...			X												
	Charlie & Margot	X														
Genre	Rom-com															
	Musical			X		X										
	Horror	X		X												
	Western	X	●		●											
	Action		●	?	●											
Friend	Tiffany	X														
	Molly					X										
	Claire			X												
	Sandra															
	Brittany	X														

The only possible way for these conditions to work together would be if the horror or the western premiered in October, which means that neither of those genres could have aired in September or November, because they must be separated by one month (per Clue 1).

Mavis Goes to the Movies		Month					Friend					Genre				
		December	November	October	September	August	Brittany	Sandra	Claire	Molly	Tiffany	Action	Western	Horror	Musical	Rom-com
Movie	Our trip...					X										
	A summer....			X												
	A talk...					X										
	Meet me...			X												
	Charlie & Margot	X														
Genre	Rom-com			X		X										
	Musical	X		X	X											
	Horror	X	X		X											
	Western			X	X											
	Action			X												
Friend	Tiffany	X														
	Molly					X										
	Claire				X											
	Sandra															
	Brittany	X														

Considering that the rom-com and the musical aired in consecutive months (per Clue 2), we can now eliminate rom-com from November and musical from September. Additionally, no matter what the final position of these two would be, there is no way they would be placed in the September and November slots (as these are not consecutive), so one of these slots must be taken by the action movie.

Now we can utilize Clue 3:

- **One movie will premiere between the musical and the movie Mavis will watch with Brittany, in some order.**

We were able to eliminate the musical from the October premiere, meaning that the movie with Brittany will not happen in August or November.

Now we can use Clue 4:

- **The rom-com will premiere after the movie Mavis wants to watch with Brittany.**

From this clue, we know the rom-com could not have aired in September, and now we are able to pin down the months for the rom-com, the musical, and the action movie, as well as the months Mavis will watch movies with Brittany and Molly, because of the information in Clue 5:

- **Molly will enjoy a movie with Mavis sometime after the musical's premiere.**

Mavis Goes to the Movies		Month					Friend					Genre				
		December	November	October	September	August	Brittany	Sandra	Claire	Molly	Tiffany	Action	Western	Horror	Musical	Rom-com
Movie	Our trip...					X										
	A summer....			X												
	A talk...					X										
	Meet me...			X												
	Charlie & Margot	X														
Genre	Rom-com	✓	X	X	X	X										
	Musical	X	✓	X	X	X										
	Horror	X	X		X											
	Western	X	X		X											
	Action	X	X	X	✓	X										
Friend	Tiffany	X				X										
	Molly	✓	X	X	X	X										
	Claire	X		X	X											
	Sandra	X				X										
	Brittany	X	X	X	✓	X										

At this point, we have significantly narrowed things down, and can apply information from the other clues.

Let's begin with Clue 6:

- **"Charlie & Margot" premieres the month before Mavis is scheduled to go to the movies with Molly.**

We know the movie night with Molly was in December, so we can locate the premiere of "Charlie & Margot".

Then Clue 7:

- **Two movies have premieres between "A Summer to Remember" and the movie Mavis will enjoy with Claire, in some order.**

This is now limited to a single working possibility.

		Month					Friend					Genre				
	Mavis Goes to the Movies	December	November	October	September	August	Brittany	Sandra	Claire	Molly	Tiffany	Action	Western	Horror	Musical	Rom-com
Movie	Our trip...		X			X										
	A summer....	X	X	X	X	✓										
	A talk...		X			X										
	Meet me...		X	X		X										
	Charlie & Margot	X	✓	X	X	X										
Genre	Rom-com	✓	X	X	X	X										
	Musical	X	✓		X	X										
	Horror	X	X		X											
	Western	X	X		X											
	Action	X	X	X	✓	X										
Friend	Tiffany	X	X		X											
	Molly	✓	X	X		X										
	Claire	X	✓	X	X	X										
	Sandra	X	X		X											
	Brittany	X	X	X	✓	X										

The movie with Sandra happened in either August or September, and "Our Trip to the Frog Pond" could not have aired in November. We also know the information in Clue 8:

- **"Our Trip to the Frog Pond" will premiere sometime after the horror movie. Either the month before or the month after "Our Trip to the Frog Pond" premieres, Mavis will see a movie with Sandra.**

That leaves September as the only working slot, which, in turn, means the horror movie aired in August and the western premiered in October. After all that, "Meet Me at Dusk" has one possibility left, allowing us to also solve the first subgrid.

	Mavis Goes to the Movies	Month					Friend					Genre				
		December	November	October	September	August	Brittany	Sandra	Claire	Molly	Tiffany	Action	Western	Horror	Musical	Rom-com
Movie	Our trip...	X	X	X	✓	X										
	A summer....	X	X	X	✓											
	A talk...	X	X	X	X	✓										
	Meet me...	✓	X	X	X											
	Charlie & Margot	X	✓	X	X	X										
Genre	Rom-com	✓	X	X	X	X										
	Musical	X	✓	X	X	X										
	Horror	X	X	X	X	✓										
	Western	X	X	✓	X	X										
	Action	X	X	X	✓	X										
Friend	Tiffany	X	X		✓											
	Molly	✓	X	X	X	X										
	Claire	X	✓	X		X										
	Sandra	X	X		✓											
	Brittany	X	X	X	✓	X										

71

Now onto our final clue:

- **Mavis will enjoy a movie with Tiffany before the premiere of "A Talk Under the Moon."**

This provides the last detail we need to complete the puzzle:

	Mavis Goes to the Movies	Month					Friend					Genre				
		December	November	October	September	August	Brittany	Sandra	Claire	Molly	Tiffany	Action	Western	Horror	Musical	Rom-com
Movie	Our trip...	X	X	X	✓	X										
	A summer....	X	X	X	X	✓										
	A talk...	X	X	✓	X	X										
	Meet me...	✓	X	X	X	X										
	Charlie & Margot	X	✓	X	X	X										
Genre	Rom-com	✓	X	X	X	X										
	Musical	✓	X	X	X	X										
	Horror	X	X	X	✓	X										
	Western	X	X	✓	X	X										
	Action	X	X	X	✓	X										
Friend	Tiffany	X	X	X	✓	X										
	Molly	✓	X	X	X	X										
	Claire	X	✓	X	X	X										
	Sandra	X	X	✓	X	X										
	Brittany	X	X	X	✓	X										

Congratulations!

You have solved your first five logic grid puzzles - and these last ones were not for the faint of heart!

Now you are ready to start solving on your own, and below you will find 25 more puzzles. The solutions are available at the end of the book.

Puzzles

At the Doctor's Office

Five doctors have their offices on the fourth floor of the medical building. Each of them has a different specialty and works with a different secretary. Offices 401 through 405 are all in a row, with office 401 being the leftmost.

1. Two things we know about Patty. First, there is one office between the one where she works and the one where Lucy works, in some order. Second, there is one office between the one where she works and Dr. Thompson's office, with Patty's office being left of Dr. Thompson's office.
2. The traumatologist's office is somewhere to the right of Lucy's office.
3. Three things we know about the neurologist. First, there is one office between the cardiologist's office and his, in some order. Second, the office where Emily works is located somewhere to the right of his. Third, his office is not number 401.
4. The endocrinologist's office is somewhere to the right of Gaby's office. Dr. Thompson's office, on the other hand, is somewhere to the left of the one where Gaby works.
5. Dr. Williams' office is located somewhere to the left of Dr. Roberts' office.
6. Dr. Jones' office is located somewhere to the left of the traumatologist's office.
7. The office where Dani works is located to the right of Dr. Stevens' office.

At the Doctor's Office

		Office					Secretary					Specialization				
At the Doctor's Office		405	404	403	402	401	Patty	Lucy	Gabby	Emily	Dani	Traumatologist	Oncologist	Neurologist	Endocrinologist	Cardiologist
Doctor	Dr. Jones															
	Dr. Roberts															
	Dr. Stevens															
	Dr. Thompson															
	Dr. Williams															
Specialization	Cardiologist											
	Endocrinologist											
	Neurologist											
	Oncologist											
	Traumatologist											
Secretary	Dani						
	Emily						
	Gabby						
	Lucy						
	Patty						

Doctor	Office	Secretary	Specialization

A Day at the Races

The international Hustlin' Hooves horse race is being held today. The first five racehorses have different colored coats, and each belongs to a rider of a different nationality. Naturally, each horse also has a unique name.

1. Two things we know about Lightning. First, he took longer to complete the race than Icarus, but less time than the chestnut. Second, he is the pinto.
2. The chestnut and the black horse finished the race in consecutive placements, in some order. The black horse finished either just before or just after Lucky Charm.
3. The Brazilian rider finished just before the Italian one.
4. The Italian finished the race sometime before Champion. He also finished just before or just after Warrior.
5. One horse finished the race between the Brazilian and the Japanese rider, in some order.
6. One horse finished the race between the white horse and the Japanese rider, in some order. The Japanese rider also took longer to finish the race than Icarus.
7. One horse finished the race between the British rider and the white horse, in some order.

A Day at the Races

A Day at the Races		Placement					Color					Nationality				
		Fifth	Fourth	Third	Second	First	White	Pinto	Gray	Chestnut	Black	Japanese	Italian	British	Brazilian	American
Horse	Champion															
	Icarus															
	Lightning															
	Lucky Charm															
	Warrior															
Nationality	American															
	Brazilian															
	British															
	Italian															
	Japanese															
Color	Black															
	Chestnut															
	Gray															
	Pinto															
	White															

Horse	Placement	Color	Nationality

The Mismatched Wedding Tables

Charlotte has decided to have a colorful and heterogeneous wedding reception. Each table has a different color tablecloth, a different type of flower for the centerpiece, and will be served a different wine. The five tables are in a row, numbered in order from left to right.

1. The table with the lavender centerpiece is somewhere right of the table being served merlot.
2. There is one table between the table with the orchid centerpiece and the table with the red tablecloth.
3. The table with the red tablecloth is somewhere to the right of the table with the white tablecloth. Speaking of the table with the white tablecloth, it's next to the table being served pinot, in some order.
4. The table with the poppy centerpiece is somewhere to the left of the one with the iris centerpiece.
5. Two things we know about the table being served champagne. First, it is next to the one being served riesling and next to the one being served pinot, in some order. Second, there is one table between the champagne table and the table with the yellow tablecloth, in some order.
6. The table being served pinot is directly to the left of the table with the poppy centerpiece.
7. The table being served sauvignon is somewhere to the left of the one with the green tablecloth.

The Mismatched Wedding Tables

The Mismatched Wedding Tables		Table					Wine					Tablecloth				
		5	4	3	2	1	Sauvignon	Riesling	Pinot	Merlot	Champagne	Yellow	White	Red	Green	Brown
Floral	Bluebells															
	Irises															
	Lavender															
	Orchids															
	Poppies															
Tablecloth	Brown															
	Green															
	Red															
	White															
	Yellow															
Wine	Champagne															
	Merlot															
	Pinot															
	Riesling															
	Sauvignon															

Table	Flowers	Wine	Tablecloth

Fantasy Adventure

There are many versions of the renowned video game Fantasy Adventure. Five versions were released between 2011 and 2015. In each version, the player has adventures in a different setting, fights with a different weapon, and plays a character with a different superpower.

1. Two things we know about the version of the game set in the countryside. First, it is older than the one set in the city. Second, there is one game version between the version set in the countryside and the version where you use a gun, in some order. By the way, the version where you use a gun is in a forest setting.
2. Two things we know about the game version where you use a knife. First, it was released either the year before or the year after the version featuring super speed. Second, it's a later version than the one with nunchucks.
3. Three things we know about the version of the game with instant healing. First, it was released either the year after or the year before the version with the seaside setting. Second, it was released before the version featuring nunchucks. Third, it was released after the version with telekinesis.
4. Two things we know about the version of the game where you use a sword. First, one version of the game was released between this version and the one where you use a gun, in some order. Second, it was released before the version with a forest setting.
5. The mountain setting was released before the seaside setting.
6. The version of the game with telepathy was released before the one with telekinesis.

Fantasy Adventure

Fantasy Adventure		Year					Weapon					Superpower				
		2015	2014	2013	2012	2011	Sword	Nunchucks	Knife	Gun	Club	Telepathy	Telekinesis	Super strength	Super speed	Instant healing
Setting	City															
	Countryside															
	Forest															
	Mountain															
	Seaside															
Superpower	Instant healing															
	Super speed															
	Super strength															
	Telekinesis															
	Telepathy															
Weapon	Club															
	Gun															
	Knife															
	Nunchucks															
	Sword															

Setting	Year	Weapon	Superpower

Locker Buddies

Five friends share a row of lockers at school, with the lowest-numbered locker on the left, and the highest-numbered locker on the right. Each of them wrote their name on a different- colored sticker on their locker, and each of their lockers has a different combination.

1. Two things we know about the locker with a gray sticker. First, there is one locker between this locker and the one with the green sticker, in some order. Second, there are two lockers between it and the one with the combination 2045, in some order.
2. There is one locker between Max's locker and the locker with the combination 2045, in some order.
3. Three things we know about Lily's locker. First, it's somewhere to the right of the locker with the 8962 combination. Second, it is next to the locker with a green sticker. Third, it is next to the locker with the white sticker, in some order.
4. Two things we know about the locker with the combination 4769. First, there is one locker between this locker and Alice's locker, in some order. Second, it is somewhere to the left of the locker with the purple sticker.
5. The locker with the black sticker is somewhere to the right of Joe's locker.
6. Max's locker is somewhere to the right of the locker with the 3561 combination.

Locker Buddies

Locker Buddies	Number					Sticker					Combination				
	125	124	123	122	121	White	Purple	Green	Gray	Black	8962	7856	4769	3561	2045
Owner Alice															
Joe															
Lily															
Maggie															
Max															
Combination 2045															
3561															
4769															
7856															
8962															
Sticker Black															
Gray															
Green															
Purple															
White															

Owner	Number	Sticker	Combination

The Art Contest

A group of local artists participated in a contest where each artist could paint any theme, but could only use two colors in their piece. Here are the top five winners and some details about their winning entries.

1. The painting in yellow and orange tones placed just in front of or just behind Gabrielle Tubbins' painting.
2. Gabrielle Tubbins' painting and the painting of a room had one painting that placed between the two of them, in some order.
3. Three things we know about Hugh Jenkins. First, his painting was awarded a higher placement than the one in green and blue tones. Second, one painting placed between Hugh Jenkins' painting and the painting in pink and purple tones, in some order. Third, his painting was beaten by the landscape painting.
4. The painting of a dog beat Jamie Miller's painting, but did not obtain third place. The dog painting also placed either just in front of or just behind the portrait.
5. Paul Spencer painted a room as his subject. His painting placed either just above or just below Jaime Miller's painting.
6. Jaime Miller placed higher than the painting in red and brown tones.

The Art Contest

The Art Contest		Place					Colors					Subject				
		5	4	3	2	1	Yellow & orange	Red & brown	Purple & pink	Green & blue	Black & white	Room	Portrait	Landscape	Dog	Abstract
Artist	Gabrielle Tubbins															
	Hugh Jenkins															
	Jamie Miller															
	Nicky Tanner															
	Paul Spencer															
Subject	Abstract															
	Dog															
	Landscape															
	Portrait															
	Room															
Colors	Black & white															
	Green & blue															
	Purple & pink															
	Red & brown															
	Yellow & orange															

Artist	Place	Colors	Subject

The Tennis Tournament

Five couples of different nationalities participated in the tennis doubles competition, finishing in different positions.

1. The Canadian couple were eliminated from the tournament earlier than Henry and his partner.
2. Two things we know about Andrew and his partner. First, they were eliminated from the tournament either just before or just after Blake and his partner. Second, they did better than Claudia and her partner.
3. Three things we know about Chris. First, he and his partner lasted longer in the tournament than Tammy and her partner. Second, he and his partner lasted longer in the tournament than the English couple. Third, one couple was eliminated from the tournament between Chris' team and the Australian team, in some order.
4. Two couples were eliminated between Edward's team and Kyra's team, in some order.
5. Edward and his partner did better in the tournament than Henry and his partner.
6. The Irish couple was eliminated from the tournament before Abby and her partner.

The Tennis Tournament

The Tennis Tournament		Position					Man					Woman				
		Winners	Runners-up	Semifinals	Quarterfinals	Round of 16	Henry	Edward	Chris	Blake	Andrew	Tammy	Kyra	Claudia	Bella	Abby
Nationality	American															
	Australian															
	Canadian															
	English															
	Irish															
Woman	Abby															
	Bella															
	Claudia															
	Kyra															
	Tammy															
Man	Andrew															
	Blake															
	Chris															
	Edward															
	Henry															

Nationality	Position	Man	Woman

Dream Logic

Claire had a very strange dream last night. She had to go through a series of five doors into five rooms, proceeding from Room 1 to Room 5, one after the other. Behind each door was a creature that asked her for something valuable and gave her a piece of battle equipment in return.

1. Claire met the unicorn sometime before she gave away the powerful amulet, but sometime after she met the gargoyle.
2. Claire was asked for a bag of sapphires sometime after she was given a sword.
3. There was one room between the room where Claire gave away the pearl necklace and the room where she was given a lance, in some order. There were two rooms between the room where Claire met the gargoyle and the room where she was given a lance, in some order.
4. The centaur was in a room somewhere before the room where she was given body armor.
5. There were two rooms between the one where Claire gave away a silk scarf and the one where she met the dragon, in some order.
6. Two things we know about the helmet. First, Claire received it in a room somewhere before the room where she received the shield. Second, she received it in some room after the room where she met a vampire.

Dream Logic

Dream Logic		Door					Valuable					Equipment				
		5	4	3	2	1	Silk scarf	Sapphires	Powerful amulet	Pearl necklace	Gold coins	Sword	Shield	Lance	Helmet	Body armor
Capture	Centaur															
	Dragon															
	Gargoyle															
	Unicorn															
	Vampire															
Equipment	Body armor															
	Helmet															
	Lance															
	Shield															
	Sword															
Valuable	Gold coins															
	Pearl necklace															
	Powerful amulet															
	Sapphires															
	Silk scarf															

Door	Capture	Valuable	Equipment

91

Neighborhood Families

There are five houses on a little street in a quaint neighborhood. Each of these houses is inhabited by a family with a son, daughter, and family pet. The houses are in numerical order, with house 871 on the far left.

1. Mittens lives next door to Shirley, in some order, and also somewhere to the right of Arianna.
2. Roy lives somewhere right of Neil, but somewhere left of Scout. Speaking of Scout, he lives somewhere right of Ted, but somewhere left of Taylor.
3. Gail lives somewhere left of Pepper, but somewhere right of Fluffy.
4. Boots lives somewhere right of Shirley.
5. Ted lives next door to Ben, in some order.

Neighborhood Families

		House					Pet					Daughter				
	Neighborhood Families	875	874	873	872	871	Scout	Pepper	Mittens	Fluffy	Boots	Taylor	Shirley	Gail	Christine	Arianna
Son	Ben															
	Neil															
	Roy															
	Sam															
	Ted															
Daughter	Arianna															
	Christine															
	Gail															
	Shirley															
	Taylor															
Pet	Boots															
	Fluffy															
	Mittens															
	Pepper															
	Scout															

Son	House	Pet	Daughter

93

The Five Safes Riddle

Five houses on a little street are inhabited by five families. The houses are numbered, starting with House 111 to the utmost left and ascending in numerical order. Each family keeps a different valuable hidden in a safe inside their home, and they each have a unique combination for their safe.

1. The family that keeps cash in their safe lives next door to the family whose safe unlocks with the code 6019, in some order.
2. The family that keeps their passports in their safe lives somewhere to the right of the safe with the 6019 combination.
3. The family that keeps the will in their safe doesn't live in the house 112,
4. Four things we know about the Andersen family. First, they live next door to the family whose safe unlocks with the code 8285, in some order. Second, there is one house between their home and the Lorrimer home, in some order. Third, the safe that opens with the code 7204 is in a house somewhere to the right of the Andersen home. Fourth, they live somewhere to the left of the family that keeps jewelry in their safe.
5. The family whose safe unlocks with the code 2371 lives somewhere to the left of both the Connors and Lorrimer families. They also live next door to the Collins family, on the left or right.
6. There is one house between the one with the safe code 2371 and the one with the safe code 6019, in some order.
7. There is one house between the Collins home and the house with the code 8285, in some order.
8. The Wilkins family lives somewhere to the right of the family that keeps jewelry in their safe.

94

The Five Safes Riddle

The Five Safes Riddle		House					Combination					Valuable				
		115	114	113	112	111	8285	7204	6019	3489	2371	Will	Stamp collection	Passports	Jewelry	Cash
Family	Andersen															
	Collins															
	Connors															
	Lorrimer															
	Wilkins															
Valuable	Cash															
	Jewelry															
	Passports															
	Stamp collection															
	Will															
Combination	2371															
	3489															
	6019															
	7204															
	8285															

Family	House	Combination	Valuable

Cruise Liner Lineup

Every month, April through August, Captivating Cruises offer cruise trips to their customers. Each cruise has a different travel itinerary, and a different family has booked the VIP suite for each one.

1. The Harris family will take their cruise sometime after the Rhine cruise has sailed.
2. The Tanner family will take their cruise either the month after or the month before the Mediterranean cruise. The Mediterranean cruise, by the way, sails sometime after the Rhine one.
3. The Paradise sails sometime after the Luxurious, and sometime after the Magnificent, but sometime before the cruise that travels the Caribbean.
4. The Queen of the Sea sails sometime before the cruise along the Rhine, but sometime after the Magnificent.
5. The Pacific cruise sails sometime before the cruise across Asia.
6. The Andrews family will take their cruise sometime before The Queen of the Sea sets sail.
7. There are two months between the month the Pacific cruise sets sail and the month the Johnson family will take their cruise, in some order.

Cruise Liner Lineup

Cruise Liner Lineup		Month					Family					Itinerary				
		August	July	June	May	April	Tanner	Johnson	Harris	Colbert	Andrews	Rhine	Pacific	Mediterranian	Caribbean	Asia
Cruise	Golden Sun															
	Luxurious															
	Magnificent															
	Paradise															
	Queen of the Sea															
Itinerary	Asia															
	Caribbean															
	Mediterranian															
	Pacific															
	Rhine															
Family	Andrews															
	Colbert															
	Harris															
	Johnson															
	Tanner															

Cruise	Month	Family	Itinerary

The Radio Mix

The local radio station started their all-music segment with five songs by different bands, from different genres.

1. The song by The Spirals was played sometime after the pop song, while the song by The White Knights was played sometime before the pop song.
2. The song by The White Knights was played either just before or just after the hip-hop song.
3. Two things we know about The Bandits. First, one song was played between the Bandits' song and the song by The Rockets, in some order. Second, the Bandits' song was played sometime after the R&B song.
4. "In My Dreams" was played sometime after the song by The Spirals.
5. The second song to be played was not soul music.
6. Two songs were played between the hip-hop song and "Flying Through the Summer Sky", in some order.
7. Three things we know about "How I was Betrayed." First, it was played sometime before the R&B song. Second, it was played either just before or just after "Flying Through the Summer Sky". Third, one song was played between "How I was Betrayed" and "My Love for You is Eternal," in some order.

The Radio Mix

The Radio Mix		Order					Genre					Band				
		5	4	3	2	1	Soul	Rock	R&B	Pop	Hip-hop	The White Knights	The Spirals	The Rockets	The Fireworks	The Bandits
Song	Flying through...															
	Hearing the wind...															
	How I was...															
	In my dreams															
	My love for you...															
Band	The Bandits															
	The Fireworks															
	The Rockets															
	The Spirals															
	The White Knights															
Genre	Hip-hop															
	Pop															
	R&B															
	Rock															
	Soul															

Song	Order	Genre	Band

Wedding Season

Many happy couples are finishing preparations for their big day! Five couples will tie the knot in successive months between April and August.

1. Four things we know about Hannah. First, she is getting married sometime after the bride who will carry a bouquet of lilies. Second, we know she's getting married sometime after Tom. Third, her wedding is sometime before the wedding with the tulip bouquet. Fourth, there is one month between her wedding month and Harry's wedding month, in some order.
2. The tulip bouquet will be created sometime before the dahlia bouquet.
3. There is one month between the month of the wedding with the lily bouquet and Charlotte's wedding month, in some order.
4. There is one month between the month of Fred's wedding and the month the daisy bouquet will be requested, in some order.
5. There are two months between Jane's wedding month and Harry's wedding month, in some order. Also, Harry's wedding is taking place before Martin will be married.
6. There are two months between Ross' wedding month and Elizabeth's wedding month, in some order.

Wedding Season

	Wedding Season	Month					Flowers					Groom				
		August	July	June	May	April	Tulips	Roses	Lilies	Daisies	Dahlias	Tom	Ross	Martin	Harry	Fred
Bride	Charlotte															
	Elizabeth															
	Hannah															
	Jane															
	Julia															
Groom	Fred															
	Harry															
	Martin															
	Ross															
	Tom															
Flowers	Dahlias															
	Daisies															
	Lilies															
	Roses															
	Tulips															

Bride	Month	Groom	Flowers

The Notebooks Riddle

In study hall, every student in the first row is working out of a notebook. Their seats are labeled in ascending order, starting with seat 1 on the left of the row. Each of their notebooks is a different color and is used for a different subject.

1. The student with the orange notebook is sitting somewhere to the left of the student studying literature. Similarly, the literature notebook belongs to a student sitting somewhere to the left of Josh.
2. The math notebook can be found somewhere to the right of Josh.
3. The chemistry notebook belongs to a student sitting somewhere to the right of the owner of the orange notebook.
4. There is one seat between the owner of the orange notebook and the owner of the blue notebook, in some order.
5. There is one seat between Stuart's seat and Lisa's seat, in some order. Lisa is also sitting in one of the seats next to the owner of the black notebook, in some order.
6. Vicky and Josh are sitting next to each other, in some order.
7. The owner of the chemistry notebook is sitting somewhere to the left of the owner of the blue notebook.
8. There is one seat between the owner of the purple notebook and the student studying the biology notebook, in some order.

The Notebooks Riddle

The Notebooks Riddle		Seat					Student					Subject				
		5	4	3	2	1	Vicky	Stuart	Sarah	Lisa	Josh	Math	Literature	History	Chemistry	Biology
Color	Black															
	Blue															
	Green															
	Orange															
	Purple															
Subject	Biology															
	Chemistry															
	History															
	Literature															
	Math															
Student	Josh															
	Lisa															
	Sarah															
	Stuart															
	Vicky															

Color	Seat	Student	Subject

Officers' Orders

Five US military officials are being deployed to new locations. Each has a different rank—corporal, sergeant, lieutenant, major, and colonel, in ascending order.

1. Four things we know about Gilman. First, he has a higher military rank than Ericson. Second, he has a higher military rank than Leo. Third, he has a higher rank than the man deploying to Atlanta. Fourth, he ranks immediately above Julian.

2. Two things we know about Julian. First, he ranks immediately above Matthews. Second, he is either two ranks above or two ranks below William.

3. The man deploying to Washington D.C. is either two ranks above or two ranks below the man deploying to Miami.

4. The man deploying to Atlanta has a higher military rank than the one deploying to Washington.

5. Two things we know about the man deploying to Boston. First, he's either two ranks below or two ranks above Butler. Second, he has a lower rank than the man deploying to New York City.

6. Matthews is either three ranks below or three ranks above Rex.

Officers' Orders

Officers' Orders	Rank					Deployment					Last Name				
	Colonel	Major	Lieutenant	Sergeant	Corporal	Washington	New York	Miami	Boston	Atlanta	Strauss	Matthews	Gilman	Ericson	Butler
Given Name Aiden															
Julian															
Leo															
Rex															
William															
Last Name Butler															
Ericson															
Gilman															
Matthews															
Strauss															
Deployment Atlanta															
Boston															
Miami															
New York															
Washington															

Given Name	Rank	Deployment	Last Name

The Birthday Prepper

Ashley is attending five different birthday celebrations, each in a different month, from October to February. Being an organized person, she already knows what present she's getting each friend, and has bought different-colored ribbons for the packages, so she can wrap them early and know which is which.

1. Four things we know about the video game. First, it will be gifted earlier than Charlie's present. Second, there is one month between the video game recipient's birthday month and the month when Ashley will deliver a present with a green ribbon, in some order. Third, the video game will be gifted one or more months earlier than the riddle book. And fourth, the video game will be gifted sometime before Kate's birthday.
2. Charlie has a birthday either the month before or the month after the birthday when the blue ribbon gift will be received.
3. Mary's birthday is before John's birthday, and there are two months between their respective birthday months.
4. The yellow ribbon will be used sometime before the novel is given as a gift.
5. Ashley will not use the red ribbon for the October birthday. The red ribbon will also be used sometime before the yellow one.
6. The green and blue ribbons will be used in consecutive months, in some order.
7. The blue ribbon will be used either just before or just after the month in which Ashley will give the album to one of her friends.

The Birthday Prepper

The Birthday Prepper		Month					Present					Ribbon				
		February	January	December	November	October	Video game	Riddle book	Novel	Football	Album	Yellow	White	Red	Green	Blue
Friend	Anne															
	Charlie															
	John															
	Kate															
	Mary															
Ribbon	Blue															
	Green															
	Red															
	White															
	Yellow															
Present	Album															
	Football															
	Novel															
	Riddle book															
	Video game															

Friend	Month	Present	Ribbon

A Night in Thought City

Thought City has five hotels and the local tourism critic and columnist, Rationality, is compiling reviews of each of them. In the reviews, Rationality identifies each hotel by number of stars, street name, and paint color.

1. Deduction Hotel has fewer stars than Brainiac Hotel.
2. Two things we know about the hotel on Geometry Lane. First, it has either two more or two less stars than the gray hotel. Second, it has either one star less or one star more than the hotel on Mathematics Road.
3. Three things we know about Riddle Hotel. First, it has more stars than Logic Hotel. Second, it has exactly one star less than the red hotel. Third, it has fewer stars than Deduction Hotel.
4. The red hotel has fewer stars than the brown hotel.
5. The red hotel has exactly one star less or one star more than Syllogism Hotel.
6. The hotel on Algebra Avenue has more stars than the white hotel.
7. The hotel on Multiplication Drive has either two fewer or two more stars than Syllogism Hotel.
8. The black hotel has either two fewer or two more stars than Syllogism Hotel.
9. The gray hotel has more stars than Logic Hotel.

A Night in Thought City

		Stars					Color					Street				
A Night in Thought City		5	4	3	2	1	White	Red	Gray	Brown	Black	Probability St.	Multiplication Dr.	Mathematics Road	Geometry Lane	Algebra Avenue
Name	Brainiac Hotel															
	Deduction Hotel															
	Logic Hotel															
	Riddle Hotel															
	Syllogism Hotel															
Street	Algebra Avenue															
	Geometry Lane															
	Mathematics Road															
	Multiplication Dr.															
	Probability St.															
Color	Black															
	Brown															
	Gray															
	Red															
	White															

Name	Stars	Color	Street

Five Good Witches

This is a tale of five good witches, living in a row of cabins in the middle of the forest. Cabin 1 is on the left of the clearing, and the cabin numbers ascend from left to right. Each witch has brewed a special potion, which has a distinct color and a specific effect.

1. Four things we know about the potion that grants sincere love. First, it is made by a witch living somewhere to the right of the witch that brews the yellow potion. Second, it is made by a witch who lives somewhere to the left of the witch that brews the memory improvement potion. Third, the witch that brews this potion lives somewhere to the right of Matilda. Fourth, there is one cabin between the cabin inhabited by the witch who brews the love potion and the witch who brews the wound- healing potion, in some order.
2. Irene lives somewhere to the left of the brewer of the super speed potion.
3. There is one cabin between Irene's cabin and the cabin of the blue potion brewer, in some order.
4. The potion with the power to heal wounds is made by a witch living somewhere to the left of the blue potion brewer.
5. Two things we know about the luck potion. First, there is one cabin between the cabin where the luck potion is prepared and the cabin where Theresa lives, in some order. Second, the witch who makes this potion lives somewhere to the right of Matilda.
6. The green potion is made by a witch living somewhere to the right of Cassandra.
7. The yellow potion and the blue potion are brewed in cabins next to each other, in some order.
8. The witch who makes the white potion lives somewhere to the right of the witch who makes the pink potion.

Five Good Witches

Five Good Witches		Cabin					Effect					Color				
		5	4	3	2	1	Super speed	Sincere love	Memory improvement	Healing wounds	Good luck	Yellow	White	Pink	Green	Blue
Name	Cassandra															
	Irene															
	Matilda															
	Milana															
	Theresa															
Color	Blue															
	Green															
	Pink															
	White															
	Yellow															
Effect	Good luck															
	Healing wounds															
	Memory improvement															
	Sincere love															
	Super speed															

Witch	Cabin	Effect	Potion color

Five Cities, Five Races

Five races were held, each in a different city. Each race was five miles longer than the last, and each had a different winner wearing a different-colored shirt.

1. Michael participated in a race somewhat longer than the Parisian race, but Alex's race was somewhat shorter than the Parisian one.
2. Three things we know about George. First, there was a 5-mile difference between the race he ran and the Parisian race. Second, there was a 10-mile difference between the race he won and the one with the blue-shirt-wearing winner. Third, there was a 10-mile difference between the race George participated in and the one with the white-shirt-wearing winner.
3. The New York race was longer than the race with the gray-shirt-wearing victor.
4. The man with the yellow shirt participated in a shorter race than the Berlin race.
5. There was a 5-mile difference between the London race and the race with the yellow-shirt-wearing winner.
6. The man in the blue shirt participated in a longer race than the Roman race.
7. There was a 5-mile difference between the Roman race and the race Alex won.
8. Albert took part in a shorter race than the man in the yellow shirt won.

Five Cities, Five Races

		Length					Runner					Shirt color				
Five Cities, Five Races		25 miles	20 miles	15 miles	10 miles	5 miles	Michael	George	David	Alex	Albert	Yellow	White	Gray	Blue	Black
City	Berlin															
	London															
	New York															
	Paris															
	Rome															
Shirt color	Black															
	Blue															
	Gray															
	White															
	Yellow															
Runner	Albert															
	Alex															
	David															
	George															
	Michael															

City	Length	Runner	Shirt color

Distributing Luggage

Five passengers are in line to leave their luggage at the airline counter. Each passenger is headed for a different destination. Also, each of their suitcases is a different color with a different- colored tag.

1. Three things we know about the passenger headed for Milan. First, they're somewhere behind the owner of the brown suitcase. Second, they're somewhere behind the owner of the suitcase with the green tag, but somewhere in front of the owner of the suitcase with a purple tag. Third, they're behind the passenger headed for São Paulo.
2. The yellow suitcase belongs to someone behind the owner of the purple tagged suitcase.
3. The passenger headed for Tokyo is in front of the one headed for Istanbul.
4. The suitcase with the orange tag belongs to someone in front of the owner of the red suitcase.
5. The suitcase with the pink tag belongs to the person either directly in front of or directly behind the owner of the yellow suitcase.
6. The passenger with the pink tagged suitcase is somewhere in front of the passenger headed for Edinburgh.
7. There are two people between the owner of the suitcase with the green tag and the owner of the white suitcase, in some order.

Distributing Luggage

Distributing Luggage		Passenger					Tag color					Suitcase				
		5	4	3	2	1	Purple	Pink	Orange	Green	Blue	Yellow	White	Red	Brown	Black
Destination	Edinburgh															
	Istanbul															
	Milan															
	Sao Paolo															
	Tokyo															
Suitcase	Black															
	Brown															
	Red															
	White															
	Yellow															
Tag color	Blue															
	Green															
	Orange															
	Pink															
	Purple															

Destination	Passenger	Tag color	Suitcase

At the Supermarket

Five men are standing in line at the supermarket. Each of them is buying one item, and each of them is wearing a distinctive item of clothing.

1. Rogers is directly in front of the man buying wine. Meanwhile, the man wearing a jacket is somewhere behind both of them.
2. Four things we know about Smith. First, he is directly in front of the man wearing a hat. Second, he is somewhere in front of the man buying eggs. Third, he is somewhere behind the man buying cookies. Fourth, there is one man between Smith and the man wearing suspenders, in some order.
3. Two things we know about the man buying milk. First, he's somewhere in front of the man buying fruit. Second, he's standing either directly in front of or directly behind the man wearing a jacket.
4. Talbott is the last of the five in line. He is also right next to Gilchrist.
5. The man wearing a flowered shirt is somewhere in front of the one wearing glasses.

At the Supermarket

		Place					Clothing					Purchase				
	At the Supermarket	5	4	3	2	1	Suspenders	Jacket	Hat	Glasses	Flowered shirt	Wine	Milk	Fruit	Eggs	Cookies
Name	Cox															
	Gilchrist															
	Rogers															
	Smith															
	Talbott															
Purchase	Cookies															
	Eggs															
	Fruit															
	Milk															
	Wine															
Clothing	Flowered shirt															
	Glasses															
	Hat															
	Jacket															
	Suspenders															

Name	Place	Clothing	Purchase

The Chocolate Bar Riddle

The local supermarket is fully stocked with delicious chocolate bars! Five different brands are available, each with a different filling and a different logo. They also vary in price—ranging from $5 to $9.

1. There is a $1 price difference between the bar with a star on its wrapping and the bar with peanuts.
2. The caramel-filled bar costs $1 more than the peanut-filled one.
3. There is a $2 price difference between the AlwaysFlavor bar and the one with a smiley face logo.
4. Three things we know about the chocolate bar with the moon logo. First, there's a $3 price difference between the moon logo bar and the heart logo bar. Second, there is a $2 price difference between this chocolate bar and the one with mint crème filling. Third, there is a $3 price difference between this chocolate bar and Mr. Sweet.
5. The Fancychocs bar does not cost $8.
6. The chocolate bar with peanuts is more expensive than Sugary's.
7. The chocolate bar with almonds costs either $1 more or $1 less than Mr. Sweet.

The Chocolate Bar Riddle

	The Chocolate Bar Riddle	Price					Logo					Filling				
		$9	$8	$7	$6	$5	Star	Smiley face	Moon	Heart	Bear	Toffee	Peanuts	Mint Crème	Caramel	Almonds
Brand	AlwaysFlavor															
	Fancychocs															
	Luke's Bars															
	Mr. Sweet															
	Sugary's															
Filling	Almonds															
	Caramel															
	Mint Crème															
	Peanuts															
	Toffee															
Logo	Bear															
	Heart															
	Moon															
	Smiley face															
	Star															

Brand	Price	Filling	Drawing

Standout Passengers

Five different people are at the airport, waiting for their flights to depart from different gates that have been numbered in ascending order from left to right, 201 being the leftmost one. Each passenger is traveling to a different destination, has a different reason for flying out, and has been flagged by the flight crew for special consideration.

1. The traveler headed for Auckland will depart from a gate adjacent to that of the vegetarian, in some order.
2. The business traveler will depart from a gate somewhere left of the gate where the vegetarian is waiting.
3. Three things we know about the flight to Rome. First, there is one gate between this one and the gate where the wedding attendant is waiting, in some order. Second, this gate is somewhere to the left of the gate where passengers leave for London. Third, this gate is somewhere right of the gate where the passenger visiting family must board their flight.
4. The Auckland visitor will depart from a gate adjacent to that of the passenger going on vacation.
5. Two things we know about the passenger traveling with a pet. First, there is one gate between this passenger's gate and the one from which the flight to London will depart, in some order. Second, there is one gate between the gate where this passenger is waiting and the gate where the passenger traveling to visit family is waiting, in some order.
6. The flight to Paris will not depart from gate 203.
7. The passenger awaiting the flight to New York will depart from the gate directly left of the gate being prepared for the disabled passenger.
8. The passenger flying first class will depart from a gate somewhere to the right of the gate from which the passenger who's traveling for school will depart.

Standout Passengers

	Standout Passengers	Gate					Reason					Special				
		205	204	203	202	201	Wedding	Vacation	Schooling	Family	Business	Vegetarian	Seeing eye pet	Nut allergy	First class	Disabled
Destination	Auckland															
	London															
	New York															
	Paris															
	Rome															
Special	Disabled															
	First class															
	Nut allergy															
	Seeing eye pet															
	Vegetarian															
Reason	Business															
	Family															
	Schooling															
	Vacation															
	Wedding															

Destination	Gate	Reason	Special

121

Feline Food Critics

Many cats live on this street, and they're very picky eaters! Each feline has a different name, breed, and food preference. The houses they live in are numbered in increasing numerical order, from left to right.

1. The cat who loves salmon lives next door to the Ragdoll. It also lives somewhere to the left of the Maine Coon.
2. Bailey lives somewhere to the right of the cat who loves shrimp, but to the left of the Abyssinian. Also, the shrimp-loving feline lives next door to Kitty.
3. Three things we know about the Persian cat. First, it lives somewhere to the right of the cat who loves beef. Second, it lives somewhere to the left of Pirate. Third, it lives next door to the cat who loves to eat liver.
4. Two things we know about the cat who loves chicken. First, it lives next door to Jester. Second, there is one house between the chicken-loving cat and the cat who loves beef, in some order.

Feline Food Critics

Feline Food Critics		House					Food					Breed				
		305	304	303	302	301	Shrimp	Salmon	Liver	Chicken	Beef	Siamese	Ragdoll	Persian	Maine Coon	Abyssinian
Name	Bailey															
	Jester															
	Kitty															
	Pirate															
	Princess															
Breed	Abyssinian															
	Maine Coon															
	Persian															
	Ragdoll															
	Siamese															
Food	Beef															
	Chicken															
	Liver															
	Salmon															
	Shrimp															

Name	House	Food	Breed

Bustling Bus Lines

The Transit City bus system has five bus lines. Each bus line is painted a different color, serves a different main avenue within the city, and shuttles passengers to a different location in the city.

1. Two things we know about the Lemming Ave. line. First, its line number is either three lesser or three greater than the number of the Patterson Ave. line. Second, it has a greater number than the line that visits the school.
2. The line that visits the planetarium has a greater number than the yellow bus line.
3. The number of the bus line that visits the zoo is one number lesser than the white bus line's number.
4. Three things we know about the red bus line. First, it has a number greater than the yellow line. Second, its line number is either two lesser or two greater than the number of the line that visits the museum. Third, its line number is one number greater than the blue line's number.
5. Two things we know about the Gelber Ave. bus line. First, it has a lesser number than the Franklin Ave. bus line. Second, its line number is either two lesser or two greater than the number of the line that you can take on Patterson Avenue.
6. The line number of the bus line that visits the cemetery is either three lesser or three greater than the number of the bus line that visits the planetarium.
7. The line that visits the museum has a greater number than the Harrow Ave. line.

Bustling Bus Lines

Bustling Bus Lines		Bus Line					Landmark					Avenue				
		5	4	3	2	1	Zoo	School	Planetarium	Museum	Cemetery	Patterson Ave.	Lemming Ave.	Harrow Ave.	Gelber Ave.	Franklin Ave.
Color	Blue															
	Green															
	Red															
	White															
	Yellow															
Avenue	Franklin Ave.															
	Gelber Ave.															
	Harrow Ave.															
	Lemming Ave.															
	Patterson Ave.															
Landmark	Cemetery															
	Museum															
	Planetarium															
	School															
	Zoo															

Color	Line	Landmark	Avenue

Solutions

At the Doctor's Office

Doctor	Office	Secretary	Specialization
Dr. Jones	402	Dani	Oncologist
Dr. Roberts	405	Emily	Endocrinologist
Dr. Stevens	401	Patty	Cardiologist
Dr. Thompson	403	Lucy	Neurologist
Dr. Williams	404	Gabby	Traumatologist

A Day at the Races

Horse	Placement	Color	Nationality
Champion	Fifth	Black	Japanese
Icarus	First	Gray	British
Lightning	Second	Pinto	American
Lucky Charm	Fourth	Chestnut	Italian
Warrior	Third	White	Brazilian

The Mismatched WeddingTables

Floral	Table	Wine	Tablecloth
Bluebells	1	Merlot	Brown
Irises	5	Riesling	Red
Lavender	2	Sauvignon	Yellow
Orchids	3	Pinot	Green
Poppies	4	Champagne	White

Fantasy Adventure

Setting	Year	Weapon	Superpower
City	2015	Knife	Super strength
Countryside	2011	Sword	Telepathy
Forest	2013	Gun	Instant healing
Mountain	2012	Club	Telekinesis
Seaside	2014	Nunchucks	Super speed

Locker Buddies

Owner	Number	Sticker	Combination
Alice	125	White	7856
Joe	121	Gray	3561
Lily	124	Purple	2045
Maggie	123	Green	4769
Max	122	Black	8962

The Art Contest

Artist	Place	Colors	Subject
Gabrielle Tubbins	3	Green & blue	Portrait
Hugh Jenkins	2	Yellow & orange	Dog
Jamie Miller	4	Purple & pink	Abstract
Nicky Tanner	1	Black & white	Landscape
Paul Spencer	5	Red & brown	Room

The Tennis Tournament

Nationality	Position	Man	Woman
American	Winners	Edward	Abby
Australian	Quarterfinals	Andrew	Kyra
Canadian	Round of 16	Blake	Claudia
English	Semifinals	Henry	Tammy
Irish	Runners-up	Chris	Bella

Dream Logic

Capture	Door	Valuable	Equipment
Centaur	3	Pearl necklace	Shield
Dragon	5	Powerful amulet	Lance
Gargoyle	2	Silk scarf	Helmet
Unicorn	4	Sapphires	Body armor
Vampire	1	Gold coins	Sword

Neighborhood Families

Son	House	Pet	Daughter
Ben	874	Scout	Christine
Neil	871	Fluffy	Arianna
Roy	872	Mittens	Gail
Sam	875	Boots	Taylor
Ted	873	Pepper	Shirley

The Five Safes Riddle

Family	House	Combination	Valuable
Andersen	112	2371	Stamp collection
Collins	111	3489	Will
Connors	113	8285	Cash
Lorrimer	114	6019	Jewelry
Wilkins	115	7204	Passports

Cruise Liner Lineup

Cruise	Month	Family	Itinerary
Golden Sun	August	Harris	Caribbean
Luxurious	June	Tanner	Rhine
Magnificent	April	Andrews	Pacific
Paradise	July	Johnson	Mediterranian
Queen of the Sea	May	Colbert	Asia

The Radio Mix

Song	Order	Genre	Band
Flying through...	4	R&B	The Spirals
Hearing the wind...	2	Rock	The White Knights
How I was...	3	Pop	The Rockets
In my dreams	5	Soul	The Bandits
My love for you...	1	Hip-hop	The Fireworks

Wedding Season

Bride	Month	Flowers	Groom
Charlotte	June	Tulips	Fred
Elizabeth	August	Daisies	Martin
Hannah	May	Roses	Ross
Jane	April	Lilies	Tom
Julia	July	Dahlias	Harry

The Notebooks Riddle

Color	Seat	Student	Subject
Black	4	Josh	Biology
Blue	3	Lisa	Literature
Green	5	Vicky	Math
Orange	1	Stuart	History
Purple	2	Sarah	Chemistry

Officers' Orders

Given Name	Rank	Deployment	Last Name
Aiden	Major	Miami	Gilman
Julian	Lieutenant	Atlanta	Butler
Leo	Sergeant	Washington	Matthews
Rex	Colonel	New York	Strauss
William	Corporal	Boston	Ericson

The Birthday Prepper

Friend	Month	Present	Ribbon
Anne	October	Football	White
Charlie	January	Album	Green
John	February	Novel	Blue
Kate	December	Riddle book	Yellow
Mary	November	Video game	Red

A Night in Thought City

Name	Stars	Color	Street
Brainiac Hotel	5	Gray	Algebra Avenue
Deduction Hotel	3	Red	Geometry Lane
Logic Hotel	1	White	Probability St.
Riddle Hotel	2	Black	Multiplication Dr.
Syllogism Hotel	4	Brown	Mathematics Road

Five Good Witches

Witch	Cabin	Effect	Color
Cassandra	4	Memory improvement	White
Irene	1	Healing wounds	Pink
Matilda	2	Super speed	Yellow
Milana	5	Good luck	Green
Theresa	3	Sincere love	Blue

Five Cities, Five Races

City	Length	Runner	Shirt Color
Berlin	25 miles	Michael	Blue
London	10 miles	Alex	Gray
New York	15 miles	George	Yellow
Paris	20 miles	David	Black
Rome	5 miles	Albert	White

Distributing Luggage

Destination	Passenger	Tag Color	Suitcase
Edinburgh	5	Blue	Yellow
Istanbul	4	Pink	White
Milan	2	Orange	Black
Sao Paolo	1	Green	Brown
Tokyo	3	Purple	Red

At the Supermarket

Name	Place	Clothing	Purchase
Cox	3	Hat	Eggs
Gilchrist	4	Suspenders	Milk
Rogers	1	Flowered shirt	Cookies
Smith	2	Glasses	Wine
Talbott	5	Jacket	Fruit

The Chocolate Bar Riddle

Brand	Price	Logo	Filling
AlwaysFlavor	$7	Star	Mint crème
Fancychocs	$9	Moon	Caramel
Luke's Bars	$8	Bear	Peanuts
Mr. Sweet	$6	Heart	Toffee
Sugary's	$5	Smiley face	Almonds

Standout Passengers

Destination	Gate	Reason	Special
Auckland	204	Business	First class
London	205	Wedding	Vegetarian
New York	201	Family	Nut allergy
Paris	202	Schooling	Disabled
Rome	203	Vacation	Seeing eye pet

Feline Food Critics

Name	House	Food	Breed
Bailey	304	Chicken	Persian
Jester	303	Shrimp	Maine Coon
Kitty	302	Beef	Ragdoll
Pirate	305	Liver	Abyssinian
Princess	301	Salmon	Siamese

Bustling Bus Lines

Color	Line	Landmark	Avenue
Blue	2	School	Harrow Ave.
Green	5	Museum	Franklin Ave.
Red	3	Zoo	Gelber Ave.
White	4	Planetarium	Lemming Ave.
Yellow	1	Cemetery	Patterson Ave.

There's More?

We will highly appreciate your honest review on Amazon. If you see any issues with the puzzles, please provide us some details so we can locate and fix the problem!

Also, feel free to take a look at our graphical logic grids:

It's all the same puzzles, but in graphical format. No more small letters and ambiguous clues!

Made in United States
Troutdale, OR
03/02/2024

18156327R00076